Automation and Manufacturing for the Built Environment

emerald
PUBLISHING

ice
Publishing

Automation and Manufacturing for the Built Environment

Steve Thompson

Published by Emerald Publishing Limited, Floor 5, Northspring, 21–23 Wellington Street, Leeds LS1 4DL.

ICE Publishing is an imprint of Emerald Publishing Limited

Other ICE Publishing titles:
Digital Twins for Smart Cities: Conceptualisation, challenges and practices
Li Wan, Timea Nochta, Junqing Tang and Jennifer Schooling.
ISBN 9780727766007
Structural Design of Buildings: Fundamentals in Design, Management and Sustainability
Feng Fu and David Richardson. ISBN 9781835495773
Additive Manufacturing for Construction
Biranchi Panda, Pshtiwan Shakor and Vittoria Laghi.
ISBN 9780727766410

A catalogue record for this book is available from the British Library

ISBN 978-1-83608-599-7

Cover photo: TierneyMJ/Shutterstock.com

Commissioning Editor: Michael Fenton
Content Development Editor: Cathy Sellars

Production Editor: Emma Sudderick
Typeset by: KnowledgeWorks Global Limited
Index created by: Madelon Nanninga

For Lucy, Freddie, Henry,
Digby and my parents.

Contents

About the author

Steve Thompson is an architect by profession and has over 25 years of experience in the construction industry, having worked across many sectors including residential, commercial, transportation and education. He has significant experience in building manufacturing, having worked for a global product manufacturer for over 10 years, where he was heavily involved in international research and development projects, product and system developments and real-world projects. He has developed several building systems and new business models to support alternative construction methods. He also has significant knowledge of the implications of digitalisation on construction and manufacturing, having been the founding chair of BIM for Manufacturers and Manufacturing, and a nominated expert on product data on international standards development. Steve has also led the development of automation technologies including regulations checking, viability assessments and generative design, and is driven by making technologies accessible to small businesses as well as large.

Glossary

Term	Description
Application programming interface	A software intermediary that enables two or more applications to communicate with each other
Automation	Where digital or physical technologies are used to remove the need for human involvement
Benefits	The measurable improvement resulting from an outcome or outcomes
Building automation system	Uses sensors and controls to monitor and adjust a building's services such as heating and lighting
Building management system	Manages a building's services such as heating and lighting. An intelligent BMS can bring together control systems and data from a number of different sources and in complex assets
Built environment sectors	Wider services including construction, manufacturing, MMC providers, professional services and asset management
Common data environment	A cloud-based location where information on construction projects is stored and accessible by all project participants with the relevant permissions
Computer-aided facilities management	A comprehensive asset management software solution that may include space management, maintenance scheduling and asset tracking
Conceptual data model	A high level, solution-agnostic description of the real-world information requirements of a business or businesses
Configurator	A digital tool that is used to create a design using an identified set of standard components or systems
Connected autonomous plant	Also known as connected automated plant, is equipment that can operate either entirely or partially without the involvement of humans
Data	Raw, unorganised facts such as individual weights, volumes or temperatures. Data can be structured (such as in spreadsheets and databases) or unstructured such as raw data from sensors
Delivery model assessment	An analytical, evidence-based approach to reach a recommendation on how a contracting authority should structure the delivery of a project or programme
Digital transformation	Business transformation enabled by digital technologies
Digitalisation	Uses data and digital technologies to improve processes. Examples include using product data to determine expected performance of a wall construction instead of calculating manually or tracking progress of a product delivery

Digitisation	Changing something from analogue to digital, for example scanning a paper document or entering product data into a spreadsheet or online tool so that the data can be accessed and used more readily
Engineered to order	The product is made to meet the specific requirements of a customer who has placed an order
FAC-1	A Framework Alliance Contract, a multiparty umbrella contract designed to integrate the activities of consultants, integrators, subcontractors, suppliers and others, and to align interests. It sits above other contracts and agreements and defines relationships and processes not covered by those contracts
Five case model	A means of developing a business case based on five dimensions: strategic, economic, commercial, financial and management
Full-time equivalent	Refers to the number of hours considered to be the normal workload of a full-time employed person
Functional requirement	A requirement that describes what a solution must do in order to fulfil a need: what a solution must achieve
Generative design	Computer-generated design solutions, usually creating many potential solutions based on a series of input parameter ranges
Golden thread	Involves keeping a digital record of crucial building information – starting from the design phase and continuing throughout the building's life cycle, and must be stored digitally
Human capital	In the context of the built environment this includes employment opportunities, skills development, health and wellbeing
Industrialised built environment	The use of physical and digital systems to deliver, maintain and reconfigure or upgrade assets through their life cycle, with very little to no abortive work in a plug-and-play fashion
Industrialised construction	The combined use of building assemblies and automation technologies to deliver virtually or totally complete assets
Industry 4.0	The Fourth Industrial Revolution. It involves a range of technologies that combine the physical and digital worlds
Information	Data that has been put into context to make it useful
Installed product data	Data relating to the product in use in the final built asset
Logical data model	Developed from a conceptual data model and solution agnostic, provides much more detail on data structures and relationships, typically in the form of relational tables
Machine learning	A form of artificial intelligence where algorithms are used to process large volumes of data and learn from it. Once taught, ML models then make predictions or identify patterns without being programmed to do so, and as such are more advanced than simple rules-based models

Made to order	The product is made following an order being placed by a specific customer
Made to stock	The product is made without a specific customer in mind, and without an order being placed
Manufacturing-led construction	Construction where manufactured systems or manufacturing processes are used and materially change the way an asset is delivered from a traditional on-site model: includes MMC, off-site construction and DFMA
Mass customisation	The benefits of mass production combined with the benefits of customisation for specific user requirements
Master data management (MDM)	Management of all data across a business, and potentially its supply chains, whether product or nonproduct information
Metadata	Data that describes other data, providing a structured reference that helps to sort and identify attributes of the information it describes
Model-based systems engineering	The formalised application of modelling to support system requirements, design, analysis, verification and validation activities, beginning in the conceptual design phase and continuing throughout development and later life cycle phases
Most economically advantageous tender	A tender which, using a cost-effectiveness approach or price/quality ratio, offers the best price (this will not necessarily be the lowest price)
Natural capital	In the context of the built environment this values the natural environment and addresses solutions to climate impacts through the life cycle of assets
Nonfunctional requirement	Describes the qualities of a solution: what it should be
Obtainable product data	Data that describes a product as a thing, and its properties, but which is application- and project-agnostic
Operational design domain	A limiter on where, when and in what conditions a CAP system can operate
Outcome-based procurement	Procurement focused on outcomes. It involves clearly defining required outcomes and then procuring a provider based on their capability to deliver against those outcomes rather than defining how an outcome should be achieved
Outcomes	The change in state or condition of the capital due to intervention activities
Outputs	Products or services created, or by-products or waste from a process
Physical data model	Developed from a logical data model, includes the solutions that will be used in operating the model and so can form the design of a database to manage all the relevant data

Platform	A group of technologies that are used as a base on which other applications, processes or technologies are developed
Premanufactured value	A measure of the value of work carried out off-site as a proportion of the total value of work on a project
Produced capital	In the built environment this is a combination of capital and operational costs, man-made assets and their efficiency and quality
Product application data	Data on the use of the product in a certain context, whether application-, sector- or project-specific
Product data sheet	A product data template that has been completed with information on a specific product or system
Product data template	A template for the consistent sharing of data on a particular product or system type
Product information management (PIM)	Centralised systems used to manage product information consistently, both internally and potentially to share with customers and other stakeholders
Product lifecycle management (PLM)	Covers the development lifecycle of products, from initial concept through to market maturity and finally retirement of a product
Product platform	A collection of assets that are shared by a number of products. Those assets are not necessarily shared physical products but can also be processes, plant, planning (knowledge) or people
Production data	Data required for the manufacture and distribution of products
Real-time location system	A system that enables the accurate 3D location of something relative to something else, for example location within a building where GPS cannot be used
Robot	Automatically controlled, reprogrammable multipurpose manipulators, programmable in three or more axes, which can be either fixed in place or fixed to a mobile platform for use in automation applications in an industrial environment
Should cost model	Pretender estimates of what a project should cost over its whole life (delivery phase plus its full design life)
Single task construction robot	Systems that execute one specific construction task
Social capital	In the built environment this refers to influence and consultation, equality and diversity as well as changes to people experience
Supply chain digital twin	A supply chain digital twin model can be used to record the real-time condition, location and state of product and service transactions
Systems engineering	An interdisciplinary approach to the requirements definition, design, integration and management of complex systems
Value	Quantifiable financial or nonfinancial worth which is important to clients and their stakeholders in the context of an intervention

Abbreviations

Term	Description
AI	artificial intelligence
API	application programming interface
BAS	building automation system
BAU	business as usual
BCIS	building cost information service
BIM	building information modelling
BMS	building management system
bSDD	buildingSMART data dictionary
CAFM	computer-aided facilities management
CAM	computer-aided manufacturing
CAP	connected autonomous plant or connected automated plant, the meaning is the same
CDE	common data environment
CIH	construction innovation hub
CIOB	Chartered Institute of Building
CLC	Construction Leadership Council
CMC	capabilities for modern construction
CPQP	construction product quality planning
DFD	design for disassembly
DFMA	design for manufacture and assembly
DLUHC	Department for Levelling Up, Housing and Communities
DMA	delivery model assessment
DoP	declaration of performance
ECI	early contractor involvement
EPC	energy performance certificate
EPD	environmental product declaration
ESG	environmental social governance
ESI	early supply chain involvement
FTE	full time equivalent
GDP	gross domestic product
GIIG	Government and Industry Interoperability Group

GIRI	Get It Right Initiative
GPR	ground penetrating radar
GUID	globally unique identifier
GVA	gross value add
IA	intelligent automation
ICE	Institution of Civil Engineers
IFC	industry foundation classes
IM	information management
ISMS	information security management system
MBSE	model based systems engineering
MDM	master data management
MEAT	most economically advantageous tender
ML	machine learning
MMC	modern methods of construction
NLP	natural language processing
NUAR	national underground asset register
ODD	operational design domain
p-DFMA	product design for manufacture and assembly
PIM	product information management
PLM	product lifecycle management
PMV	premanufactured value
QFD	quality function deployment
RIBA	Royal Institute of British Architects
RPA	robotic process automation
RTLS	real-time location system
SCDT	supply chain digital twin
SCM	should cost models
SCOR	supply chain operations reference model
SDG	sustainable development goals
SFCECA	standard form of civil engineering cost analysis
SIC	Standard Industrial Classification
SKU	stock keeping unit
SME	small- and medium-sized enterprise
STCR	single task construction robot

Section 1

Setting the scene

emerald
PUBLISHING

ice
Publishing

Steve Thompson
ISBN 978-1-83608-599-7
https://doi.org/10.1108/978-1-83608-598-020241001

Chapter 1
Introduction

1.1.   The challenge

The world in the first half of the 21st century holds significant promise for the future in terms of technology development, living standards and health and wellbeing, arguably more so than at any other time throughout history. However, there are also real and significant global challenges ahead. One of the questions this book looks to answer is how can new approaches to delivering and maintaining built assets help address the pressing challenges faced, and provide a sustainable built environment for all? It aims to show how, with the right balance of technologies and manufacturing-led approaches, project and wider societal value are not mutually exclusive; they can be aligned and addressed simultaneously.

One challenge is the formidable expansion rate of the world's population, combined with increasing expectations for the built environment as urbanisation increases. Globally, the world's urban population is expected to grow by 2.3 bn between now and 2050 (UN DESA, 2019). That means the equivalent of requiring over 2 300 new apartment buildings being delivered every day between now and 2050, each housing 100 people. When necessary supporting infrastructure and services are added to that, it is estimated that globally over 230 bn m^2 of additional built floor area will need to be added between now and 2060 to keep up with demand – the equivalent of delivering the built area of Japan every year (zu Ermgassen *et al.*, 2022). All of this means that the need for new assets has never been greater.

Another challenge facing the built environment is quality. The impact of poor quality, inefficient built assets is covered in Chapter 2, but a key issue is that carbon emissions from the built environment are significant. The demand for meeting space requirements seems at odds with the need to reduce overall emissions: this is not only a case of balancing increasing demand with delivery but also about delivering better performing assets. Studies have shown that part of the problem in reducing emissions is that many buildings do not perform as they were intended to, emitting an average of 3.8 times the emissions that they were designed to (Innovate UK, 2016). In 2019, global carbon dioxide emissions from the operation of buildings reached 10 GtCO$_2$ per year, which equates to 28% of all emissions. When the emissions from the construction industry were added, the total energy-related emissions from construction and the built environment reached 38% of the total (UNEP, 2020).

To help minimise climate change, the UK government's Climate Change Act 2008 has committed the UK to achieving net zero greenhouse gas emissions by 2050 (HMG, 2019), with an interim target of achieving 78% lower emissions than in 1990 by 2035. To achieve anything like these targets will require a significant change in the way the world operates – for example, a greater than 6% reduction in emissions from buildings will be required every single year to 2030 to contribute to this goal (UNEP, 2020), a target which is not currently being met. Business as usual projections suggest that if the UK continues its current path, the UK will only achieve a 60% reduction in emissions from 1990 levels by 2050, falling significantly short of its net zero target (UKGBC, 2021). Therefore, if the UK is to meet its 2050 targets, it is crucial that emissions from the built environment are reduced to zero. So, assets also need to perform better.

Put simply, the construction industry exists to deliver and maintain the built environment, which itself supports the communities and individuals it serves. But what does the construction industry of the early 2020s look like, and how is it likely to change between now and 2050? Chapter 3 covers today's construction industry in some detail but, to simplify the story, many buildings and infrastructure assets that are built today would have been constructed in a very similar way many decades ago. Brick and block are still the most common materials used in constructing new homes, and on-site productivity has decreased since the early 1970s (ONS, 2021). At the same time, productivity across all industries has increased by 49% and, looking forward, material use across all industries is forecast to more than double by 2060, with over a third of that rise being used in construction, which is unsustainable. So, as there will be an ever-increasing demand for new or refurbished assets, the construction industry needs to deliver significantly more and more efficient assets, with fewer resources, and with an ever-increasing skills shortage. This simply cannot be achieved using the same approaches that have been used in the past, and Chapter 3 will describe some of the green shoots of change that have begun to develop over recent years. Further chapters will then describe how the carefully considered application of automation and manufacturing-led construction can support a move to an industrialised construction industry (how assets are delivered), and even potentially an industrialised built environment (how assets are maintained, operated and upgraded). However, there are risks involved if the role these technologies play is not carefully considered from the outset, so this book will describe important considerations and decisions that need to be made to succeed. Just using technology to deliver assets quicker is not the answer to the world's challenges.

1.2.   The role of technology and manufacturing

The last two decades have seen rapid developments in technologies that have changed the way we live our lives. The first iPhone was launched in 2007, and now the equivalent of over 78% of the world's population has a smart phone (Statista, 2022). Over 90% of the world's data has been produced in the last two years, and every 18 months to two years the amount of data doubles. Since 2018 there have been more devices connected to the internet than there are people on the planet. But what does all this technology development mean to the construction industry and the built environment? Yes, there are an increasing number of smart devices in our homes, such as learning thermostats, robotic vacuum cleaners and voice-controlled devices, but the construction industry is often described as one of the least digitalised industries and one of the least productive. The next few chapters will discuss whether this is a fair assessment and how the industry can improve. So how can technology help? There are many new terms being used such as industrialisation, the Fourth Industrial Revolution, Industry 4.0 or even Construction 4.0, and new approaches or methodologies, such as BIM (building information modelling), IM (information modelling), automation, AI (artificial intelligence), IA (intelligent automation), blockchain, generative design, offsite manufacturing, MMC (modern methods of construction) and product platforms, but what do they all mean, and are they likely to improve the industry?

Box 1.   Industry 4.0

Industry 4.0 is another term for what is known as the Fourth Industrial Revolution.

The First Industrial Revolution was brought about using water and steam power to drive the mechanisation of production in factories in the late 18th century. It changed the economy from an agriculture- to a production-led economy, bringing together disparate workers in the textiles industries into large cotton mills in towns and cities. Over time it led to increases in output, in wages and in quality of life for the general population. The Second Industrial Revolution was driven by the use of electricity and mass production in the late 19th and early 20th centuries, and the Third Industrial Revolution from the last third of the 20th century to today has developed the use of computers, digitalisation and electronics to support mass customisation (where the benefits of mass production can be combined with the ability to customise output instead of mass-producing identical products). As with the first two industrial revolutions, this promises to improve wages and living standards for many. However, while there is little doubt the Third Industrial Revolution is still being experienced, many believe that society is also now at the beginning of a Fourth Industrial Revolution, or Industry 4.0, running alongside the third. Building on the developments of the third, it is characterised by a fusion of technologies that is blurring the lines between the physical and digital worlds and promises new services and business models driven by the creation of cyber-physical systems (systems that include the integration of digital systems that can control physical systems and vice versa). Common principles included in the vision of Industry 4.0 (iscoop, 2022) include

- interoperability between systems, both physical and digital
- information transparency
- decentralisation and autonomous decisions
- real-time capability, data and decision making
- service orientation (human–machine interaction)
- modularity
- individualisation or personalisation of products with highly flexible production
- customer-centric.

Boston Consulting Group (BCG) refers to Industry 4.0 as the convergence of the following nine technologies (BCG, 2016)

- advanced robotics
- additive manufacturing
- augmented reality
- simulation
- horizontal/vertical integration
- industrial internet
- cloud
- cybersecurity
- big data and analytics.

While the first three industrial revolutions were typically focused on the factory and production systems, Industry 4.0 extends much further into other areas of our lives and industries, which is where terms such as Construction 4.0 fit.

This book touches on each of them, but more specifically looks at what automation and manufacturing-led construction mean in the context of the built environment and what practical steps can be made to take advantage of technologies. It will look to describe how they are currently being implemented, with real-world examples, and what impact they are likely to have on a future built environment. Critical to all these technologies, however, is data; they all rely on it to function. They depend on how data is captured, structured, exchanged and used, and this will be covered in more detail in Chapter 10.

The use of data and technology can have significant impacts on individuals and whole communities alike in other ways too. A prime example of this is Engel's pause; a term used to describe one of the impacts of the First Industrial Revolution. While mechanisation was great for factory owners and increased per capita GDP, it took generations before the working classes saw increases in wages as a result, leading to increased social inequality. There is a commonly held fear that the same may occur with Industry 4.0; that robots may steal the jobs of humans, that automation may replace entire occupations and it may take generations for new roles to be created for those that are displaced. Chapter 22 looks at the future of work and the implications of automation and manufacturing-led construction in the short, medium and long term.

While there is potential risk associated with jumping into using new technology and data to change the way we live and work, this book offers guidance to ensure that automation technologies can realise the significant benefits to clients, industry and society that they promise, on projects large and small. To mitigate potential risks, it is crucial to carefully define and plan future strategies for technology adoption, and Chapters 7 and 8 outline how to enable such change, including 'no regret' decisions that will provide solid foundations for the future.

It is important to highlight that Industry 4.0 is still a long way off becoming a widely held reality and is generally still only a vision for the future. To implement it fully requires systemic change across industries, but instead of waiting for that to become possible while the world's challenges continue to deepen, there are aspects which can be delivered today without being a detriment to future applications. Therefore, this book does not cover Industry 4.0 in its entirety but describes some aspects of it that can be realised today, with real examples and benefits. Different levels of adoption of automation will then be considered as part of the future scenarios described in Chapter 23.

1.3.   The way forward

Making the wrong decisions today on what we build and how we operate the built environment could lock the country into emissions in the future that mean national targets cannot be met, but can also lead to the wrong assets in the wrong location. So careful consideration needs to be given, and simply continuing to build more of the same is unlikely to be the answer. Equally, the wrong application of technologies can be inefficient, expensive and disruptive. Figure 1.1 illustrates part of the challenge; in delivering a built environment that meets society's needs, it is important to ensure that the construction industry is not just delivering things well, but also delivering the right things. Construction is historically a project-based industry, with a focus purely on meeting a specific project's objectives, whereas there is a gradual move in recent years to incorporate broader social value using tools such as the Construction Innovation Hub's Value Toolkit or the Government's

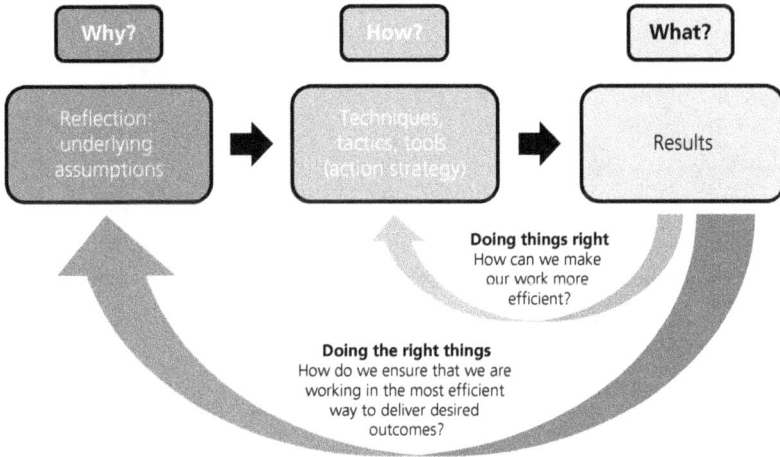

Figure 1.1 Doing the right things, not just doing things right (Author's own)

Green Book (HMT, 2022). Care needs to be given to clearly defining what is needed through the lifecycle of an asset and potentially as part of a larger portfolio of assets.

In the UK, for several years there has been a notional target from successive governments of delivering 300 000 new homes per year. However, this target has never been met. Importantly however, if the industry were to deliver that many additional homes, their delivery and subsequent use would single-handedly consume England's total cumulative carbon budget between now and 2050 (zu Ermgassen *et al*. 2022). That is not to say that more homes should not be built, but that there needs to be consideration on what and where homes need to be delivered.

On top of the housing need, there is a need for supporting infrastructure to support communities, such as schools, transport, hospitals and assets supporting employment. Approximately 80% of the UK's building stock in 2050 already exists, so it is important to recognise the critical role that our existing assets need to play; it is not simply a challenge to build better new buildings but also to maximise the use of what already exists. By extending the life of existing assets, the overall carbon footprint of the built environment can be reduced as the need to replenish stock is reduced.

To coherently describe how to get from where the industry is now to one which is more optimal to meet the challenges of today and tomorrow, Figure 1.2 illustrates five key stages to go through. In this gradual approach, each stage builds on the previous one and adds more value. If you systemise before clearly defining what is needed, you run the risk of delivering the wrong things quicker, locking in problems for the future and limiting future applicability of systems. If you automate before you systemise, you limit value in the short and long term.

The core of this book is structured into sections relating to each of the steps to achieve maximum value; however, the intention is that each chapter can also be used in isolation to describe different aspects of automation or manufacturing-led approaches. To improve an industry requires systemic change, not just changes in small pockets of the industry. However, it will not usually be the case that the whole supply chain can be steered in a new direction together, so it remains

Figure 1.2 Stages for implementing automation and manufacturing-led delivery (Author's own)

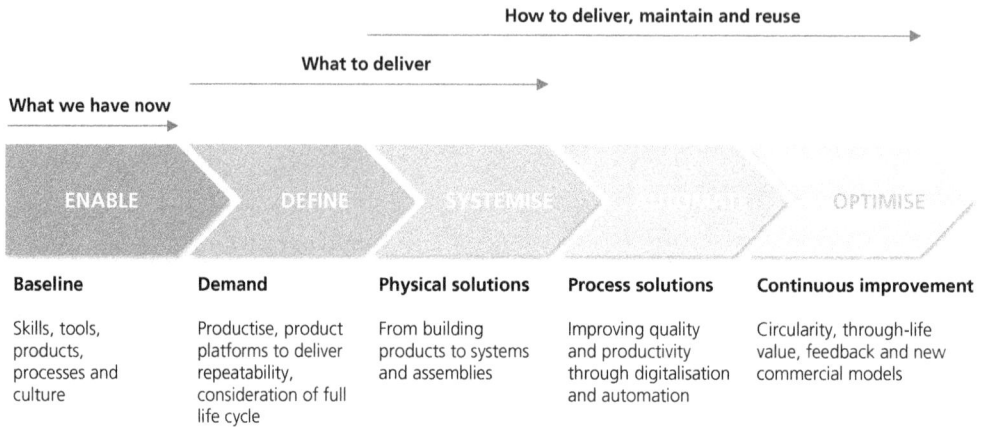

Baseline	Demand	Physical solutions	Process solutions	Continuous improvement
Skills, tools, products, processes and culture	Productise, product platforms to deliver repeatability, consideration of full life cycle	From building products to systems and assemblies	Improving quality and productivity through digitalisation and automation	Circularity, through-life value, feedback and new commercial models

important to highlight the benefits that can still be achieved when working in more focused areas with a similar vision.

The book focuses on a time frame up to 2035 to illustrate the potential of automation and manufacturing-led construction, as within that timeframe significant change is expected, and indeed needs to occur if the industry is to meet its medium- to long-term targets described in this chapter. 2035 sits 5 years after the deadline to achieve the UN's Sustainable Development Goals, and 15 years before the UK Government's net zero targets, so the book presents three scenarios for the construction industry in 2035, which will incorporate different levels of industrialisation and automation.

In continuing to set the scene, Chapter 2 looks at the built environment in more detail to describe the situation today, and what future demands are likely to be to make sure that real, enduring value is delivered through the built environment. Chapter 3 then looks at the construction industry today, including its structure from small- and medium-sized enterprises (SMEs) to large multinationals. It highlights how in recent years parts of the industry have begun to change to a value-based approach, considering the whole lifecycle of an asset, and as a result focusing more on refurbishment of existing structures than has previously been the case.

Chapter 4 describes what is meant by automation and manufacturing in the context of the construction industry and how they can act in unison to deliver greater value through a more responsive built environment.

Chapter 5 provides lessons that can be learned from other industries to support transformation of the construction industry, such as the use of digital twins to provide real-time feedback and different supply chain and value-driven industrialisation.

The last chapter, Chapter 6, in the *Setting the scene* section outlines the modelling framework that will be used throughout the book to describe and quantify the potential impact of automation and manufacturing-led construction through the lifecycle of an asset.

The *Enable* section begins with Chapter 7, which provides an overview of how the industry can prepare for change, driven by through-life value, optimising asset lifecycles and a manufacturing-led approach. Chapter 8 then presents a number of no-regret decisions; decisions that can be made and other considerations made today that are likely to add value in the future, whichever route the industry takes in the years ahead.

Chapter 9 looks at connectivity and how it plays a crucial role in pulling together a number of different technologies and approaches to maximise the value that they can add.

Chapter 10 focuses on a key aspect of future success, which is product and service data, and how the structuring of these data now will provide a solid foundation to add significant value in the future.

Section 3 *Define* is crucial to the success of new approaches to construction and the built environment and focuses on defining what needs to be delivered by the construction industry going forward, describing tools to define the value required through the lifecycle of assets. Chapter 11 focuses on defining the need, Chapter 12 on requirements management and Chapter 13 on describing what product platforms are, why they are important and how and where they can be used.

Section 4 *Systemise* focuses on systemising construction and covers different premanufacturing and integration models and how they can be combined with automation technologies. It looks into the application of integration models at different scales, but also implications through the lifecycle of assets, with real-world examples.

Section 5 *Automate* focuses on automation technologies, starting with a timeline forecasting when different technologies are likely to become widely available, with an overview of different technologies and how they can be applied at different stages through the lifecycle of an asset, from requirements management to automation during the use of an asset. The focus is on how value can be delivered through the use of the technologies being described, with real-world examples.

Section 6 *Optimise* looks at how the built environment and construction industry can be optimised for the future, including how it can support a circular economy and what the future of work looks like with increased automation and manufacturing-led approaches.

Finally, Chapter 23 presents four future scenarios for the construction industry in 2035 based on different levels of adoption of automation and manufacturing-led construction technologies.

REFERENCES

Boston Consulting Group (BCG) (2016) https://www.slideshare.net/TheBostonConsultingGroup/sprinting-to-value-in-industry-40 (accessed 18/07/2024).

Construction Innovation Hub (2020) *Value Toolkit*. Construction Innovation Hub, UK. https://constructioninnovationhub.org.uk/our-projects-and-impact/value-toolkit (accessed 18/07/2024).

HMG (Her Majesty's Government) (2019) *Climate Change Act 2008 (2050 Target Amendment) Order 2019*. The Stationery Office, London, UK.

HMT (Her Majesty'sTreasury) (2022) *The Green Book – Central Government Guidance on Appraisal and Evaluation*. The Stationery Office, London, UK.

Innovate UK (2016) *Building Performance Evaluation Programme: Findings from Non-Domestic Projects*. Innovate UK, Swindon, UK.

iscoop (2022) https://www.i-scoop.eu/industry-4-0/ (accessed 18/11/2023).

ONS (2021) (Office for National Statistics) https://www.ons.gov.uk/economy/economicout-putandproductivity/productivitymeasures/articles/productivityintheconstructionindustryuk 2021/2021-10-19 (accessed 05/10/2022).

Statista (2022) https://www.statista.com/ (accessed 05/10/2022).

UKGBC (UK Green Building Council) (2021) *Net Zero Whole Life Carbon Roadmap: A Pathway to Net Zero for the UK Built Environment*. UKGBC, London, UK.

UNDESA (United Nations Department of Economic and Social Affairs) (2019) *World Urbanization Prospects, The 2018 Revision*. UNDESA, New York, NY, USA.

UNEP (United Nations Environment Programme) (2020) *2020 Global Status Report for Buildings and Construction: Towards a Zero-Emissions, Efficient and Resilient Buildings*. UNEP, Nairobi, Kenya.

zu Ermgassen SO, Drewniok MP, Bull JW *et al.* (2022) A home for all within planetary boundaries: Pathways for meeting England's housing needs without transgressing national climate and biodiversity goals. *Ecological Economics* **201**: 107562.

emerald PUBLISHING   ice

Steve Thompson
ISBN 978-1-83608-599-7
https://doi.org/10.1108/978-1-83608-598-020241002

Chapter 2
The built environment

2.1.   Introduction

The built environment is all around humankind; it provides the context for people's lives and the environment that they live them in. It consists not only of all man-made structures collectively but also of the spaces in between. It is where people live, work and play every day of their lives; on a good day it can be the first home someone buys or the venue to celebrate a special occasion, or on a bad day the hospital buildings that enable much needed services to provide support. The built environment is effectively a complex system of systems developed independently to support societies through time and across the world, wherever they are. It incorporates both connected and unconnected assets and, importantly, as described in Figure 2.1 the built environment is not only found in urban areas but includes anywhere it is needed, from cities to rural communities and all the infrastructure in between.

The built environment is often considered at the scale of a building or buildings, or greater. However, as future chapters will show as they look in detail at the potential for systemisation and

Figure 2.1 The natural and built environments and levels of granularity (Author's own)

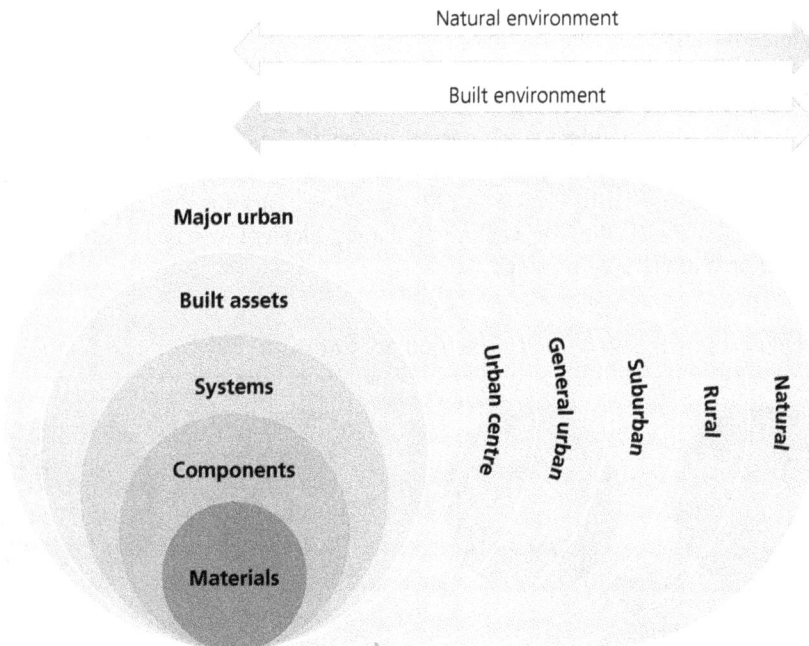

enabling a circular economy, it is crucial to consider the built environment at different levels of granularity, also highlighted in Figure 2.1. An environment cannot be created without the materials, products and systems that it comprises, and as assets move through the stages of their life cycle, different levels of granularity come in and out of focus, as illustrated in Figure 4.2 in Chapter 4. During the initial design phase it is likely to be the built asset that is the primary focus, and the spaces that it needs to incorporate. However, depending on the delivery methods chosen, products and systems are likely to come to the fore at the technical design phase or earlier and, certainly towards the end of the asset's life, products and systems will again come into focus as they are maintained, dismantled, reused or demolished.

When it comes to the reuse of assets, the preference is clearly to reuse assets at their most complete level possible (for example, extending the life of a built asset reduces the impact of its embodied carbon, whereas separating and reusing components still increases the carbon used in the process of extraction and reinstallation). This is discussed in more detail in Chapter 21.

This raises a further perspective, which is that of time. The built environment is usually considered in its completed state, with potential interventions through the delivery or removal of assets. Whereas most of the book will focus on the interventions that pull products, systems and processes together to deliver new or changed assets, it is this completed state that will be the focus of this chapter. The completed state is the output delivered through interventions over time to support the delivery of services to society (whether to individual clients or wider communities) which provide desired outcomes, as illustrated in Figure 2.2.

As mentioned in Chapter 1, the majority of the building stock that will exist in 2050 has already been built. While the assets still to be built require some delivering, later chapters will also look at how automation and manufacturing-led approaches can support the refurbishment, improvement or management of existing assets as well as delivering new. Focusing on delivering new assets alone would not address the challenges faced, nor would it demonstrate the full potential for automation and manufacturing-led construction.

2.2.   The existing building stock

So, what does the existing building stock consist of today? To give some indication, Figure 2.3 provides a breakdown of the existing stock in England and Wales by area, based on the Building Energy Efficiency Survey (BEIS, 2016) and housing stock data for the same period (DLUHC, 2022a). As Figure 2.3 shows, the overwhelmingly largest element by some distance is residential, with 70% of the area of the existing stock.

Figure 2.2 Industry output, services and value (Author's own)

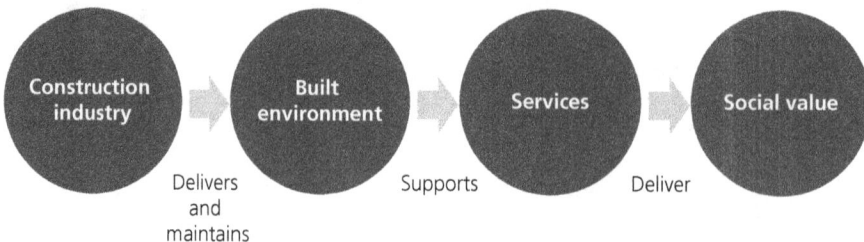

Figure 2.3 England and Wales building stock by sector 2014–15 (Source: BEIS, 2016; DLUHC, 2022a)

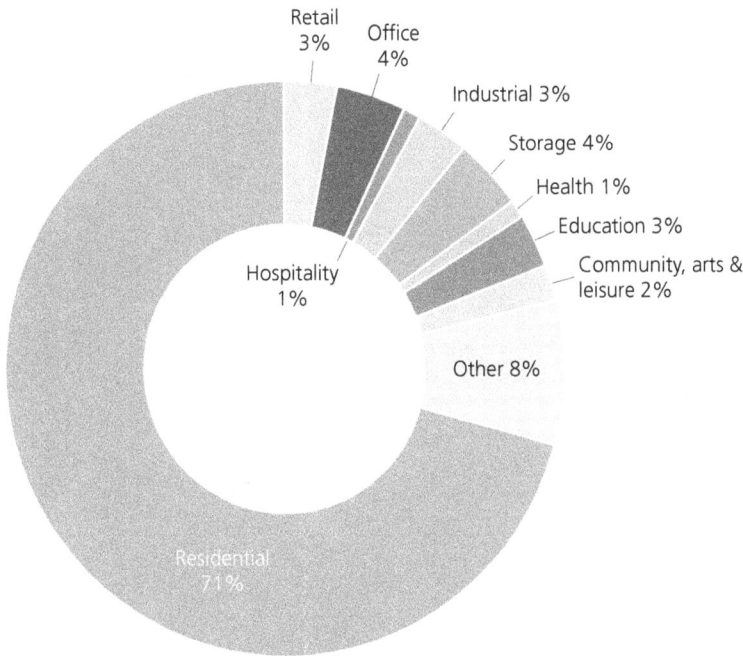

The Building Energy Efficiency Survey (BEES) (BEIS, 2016) provides the most complete picture to date of what the nonresidential building stock comprises of, how it performs today and how it could perform if potential improvements were made in the future. Figure 2.4 illustrates a further breakdown of building types by area.

While Figures 2.3 and 2.4 present useful data on the makeup of the current building stock, to consider the potential impact of new technologies or methodologies for delivering interventions, it is important to understand how they can work across traditional sector boundaries – for example, learning from delivering schools can be shared with those delivering offices or hospitals. To enable this to happen, high-level analysis has been carried out on the similarities between different built asset types, using the criteria for the entities platform described in more detail in Chapter 13. The criteria can be summarised as

- form – the complexity of the asset shape in both plan and section, including variety of space types and size
- function – complexity or level of variation in functional requirements
- performance – level of performance required for spaces and elements
- specification – level of quality required
- amenity – degree to which the public use or visit an asset
- brand – importance of brand identity.

Using this approach, a wide range of asset types has been analysed which fall into each of the sectors identified in Figure 2.3 to provide an indication of the relative importance of each criterion to

Figure 2.4 England and Wales nonresidential building stock by sector (2014–15) (Source: BEIS, 2016)

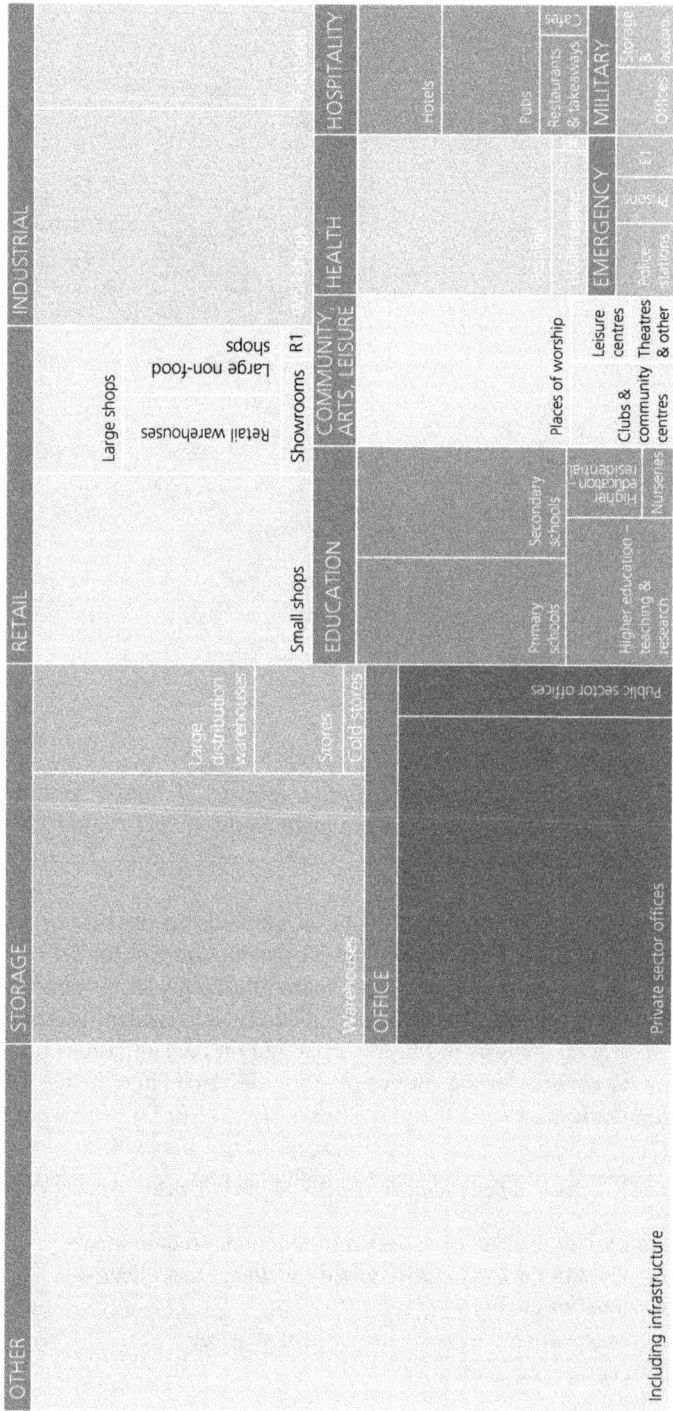

OTHER

STORAGE

Warehouses

OFFICE

Private sector offices

Public sector offices

Including infrastructure

RETAIL

Large distribution warehouses

Stores

Cold stores

Small shops

Large shops

Retail warehouses

Showrooms R1

Large non-food shops

EDUCATION

Primary schools

Secondary schools

Higher education – teaching & research

Higher education – residential

Nurseries

COMMUNITY, ARTS, LEISURE

Clubs & community centres

Leisure centres

Theatres & other

Places of worship

INDUSTRIAL

HEALTH

Hospitals

H1

HOSPITALITY

Hotels

Pubs

Restaurants & takeaways

Cafes

EMERGENCY

Police stations

E1

MILITARY

Offices

Storage & accom.

R1 – Hairdressers, H1 – Nursing homes, E1 – Fire stations & other

Figure 2.5 Building stock entities profiles (Author's own)

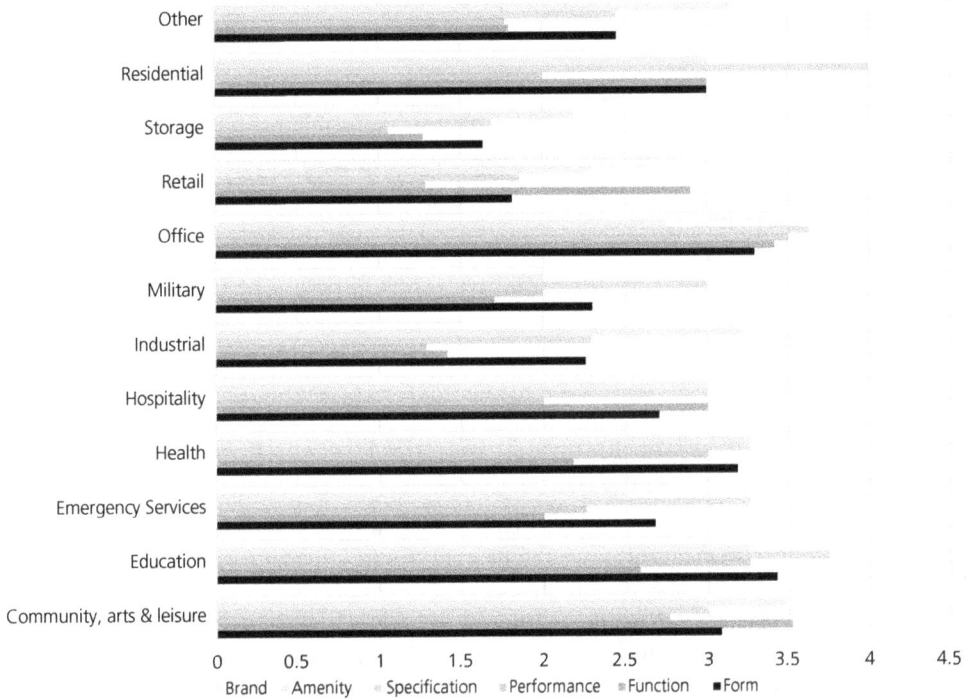

assets in a particular sector. This provides a more useful perspective on the varieties and similarities between different asset types than is traditionally considered, which comes into use when looking at alternative methods of delivering and managing assets across traditional sector boundaries. The average scores for each criterion within the different sectors are presented in Figure 2.5, but more detail can be found in Chapter 13.

Focusing now on the residential sector, there are currently more homes in England and Wales than there are households, but some of these homes are second homes and investment properties, or in the wrong location to meet the needs of households. As a result, the UK is currently experiencing a housing crisis: a chronic shortage of affordable housing or rental properties of the right type in the right place. For example, since 2000 only around 125 000 new-build homes have been built that are aimed at the retirement market. Over the same period, over 700 000 people have turned 65 (Mayhew, 2020). So, in the short to medium term there is a real need to increase access to housing that meets the requirements of its occupants. Mayhew (2020) estimates that if people lived in homes more suited to their needs, over 50 000 fewer homes would need to be built every year. This suggests that just building more homes is not the answer to the challenges ahead, even if they were not to have an impact on carbon emissions.

Figure 2.6 illustrates the mix of homes in England by their age and suggests that the average life expectancy of homes is far greater than the 60 years usually referenced. In 2020, over 52% of all homes were over 60 years old (DLUHC, 2022a), and it is important to understand the implications of this, not only on the type, amount and useability of the spaces they create over time, but also on the energy efficiency of homes.

Figure 2.6 England dwelling stock by age (2019) (Source: DLUHC, 2022a)

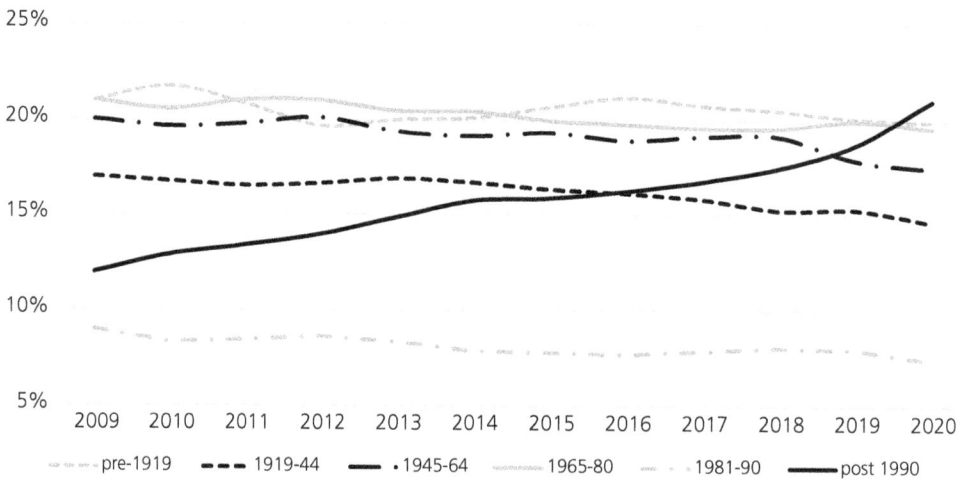

Put simply, the UK cannot achieve its zero carbon targets by just building new, more efficient assets; significant improvements also need to be made to the existing stock. In 2020, only about 10% of homes had an EPC (energy performance certificate) rating of A or B (UNEP, 2020), representing a significant challenge to get the existing stock to meet the Government's net zero carbon target by 2050. With a housing stock of over 24.8 m in England alone, and the UK Government's target to make all homes achieve an EPC rating of C or above by 2035, there needs to be on average just over a million dwellings upgraded every year for performance reasons alone[1]. To deliver such a volume of upgrades requires a drastically different approach to how homes have been upgraded in the past. However, reaching the 2035 target will not, on its own, enable the country to get to the 2050 net zero carbon target. To achieve this, improvements to the existing stock need to go beyond achieving an EPC of C for all properties; analysis suggests that to get the existing stock to perform as well as it can reasonably be expected to perform by 2050 will mean refurbishing over 73% of the existing stock – that's over 670 000 homes per year between now and 2050, plus the refurbishment of any new-build homes from today onwards that are not built to be zero carbon ready[2]. To this end, the Construction Leadership Council published their National Retrofit Strategy in 2021 (CLC, 2021), with a vision to retrofit almost all UK homes over the next 20 years.

One final aspect of the residential stock that needs to be considered before moving on to the wider built environment is that of health and safety. In the wake of the Grenfell tragedy in 2016, the Building Safety Act 2022 (HMG, 2022) is rightly likely to have a lasting impact on the built environment and the way buildings are delivered and maintained; that will be covered in more detail in the *Enable* section of this book and relies heavily on better sharing of information to deliver better informed decisions. However, the implications of poor-quality housing on the health and safety of residents are broader than the impacts highlighted in such tragedies. Figure 2.7 presents some of the results from BRE Group (2021) on the cost of poor-quality housing using England Housing Survey data and NHS treatment costs over a 25-year period.

---

[1] In 2021, 43.53% of homes in England had an EPC rating of C or above. There was a total dwelling stock of 24 873 000 homes, meaning that 14 045 783 homes needed to be upgraded over a 14-year period to achieve the 2035 target.
[2] The author carried out analysis of 330 619 homes which had an EPC carried out in 2021 (DLUHC, 2022b).

Figure 2.7 Cost of poor quality housing in England (2018) over a 25-year period (Source: BRE Group, 2021)

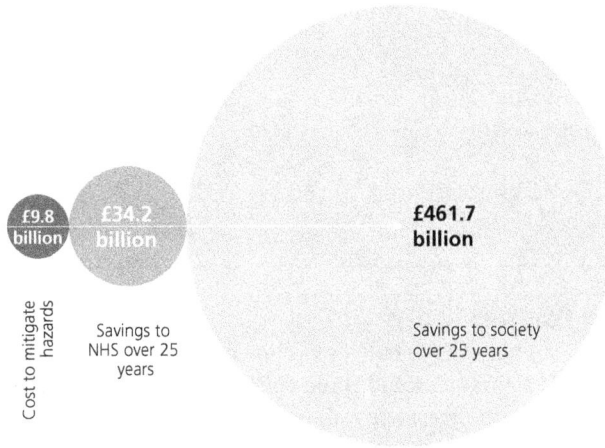

£9.8 billion

£34.2 billion

£461.7 billion

Cost to mitigate hazards

Savings to NHS over 25 years

Savings to society over 25 years

The figure shows that a relatively small spend on mitigating hazards in buildings (including incorporating better insulation and heating) can have a significant impact on the cost to society that can result from poor-quality housing. The World Health Organization suggest that globally 38 200 deaths occur every year because of cold homes (WHO, 2011), which again suggests that the operation of the built environment, and the services it provides, need to be considered when designing or delivering built assets. So, refurbishment of homes and other built assets is likely to play a much greater role in the construction industry's future work than it has in the past and, as the energy crisis in Europe in 2022 has highlighted, reducing energy bills by improving the efficiency of homes is likely to be an important factor in reducing fuel poverty in the coming decades.

Moving back to the built environment as a whole, it accounts for the largest share of global energy consumption (35% for building construction and operations in 2019), yet at the same time, for every $37 spent on conventional construction activities, only $1 is spent on energy efficiency improvements (UNEP, 2020). Figure 2.8 shows the significant difference that could be made if realistic energy efficiency improvements were made to the existing building stock in England[3].

It is increasingly clear that unless more focus is put on improving the existing stock, UK and global emissions targets in the coming decades will not be met, but there is a real question on how improvements to assets will be paid for. The necessary improvements to make the savings illustrated in Figure 2.8 to the existing stock have a payback period of between 4.8 and 17.7 years, dependent on sector (BEIS, 2016). For the residential sector, the average payback period is over 22.6 years (DLUHC 2022b), so is unlikely to be viable on pure cost savings alone. If improvements aren't started soon, however, it will become increasingly impossible to meet our evolving needs, and this is where the impact on society as a whole needs to be considered alongside individual improvement projects.

---

[3] The nonresidential improvements were identified in the Building Energy Efficiency Survey (BEIS, 2016). The residential improvements were identified in individual EPC assessments (DLUHC, 2022b). For the purposes of this analysis, over 300 000 EPC certificates were assessed.

Figure 2.8 Energy use reduction following improvements to existing stock (Source: BEIS, 2016; DLUHC, 2022b)

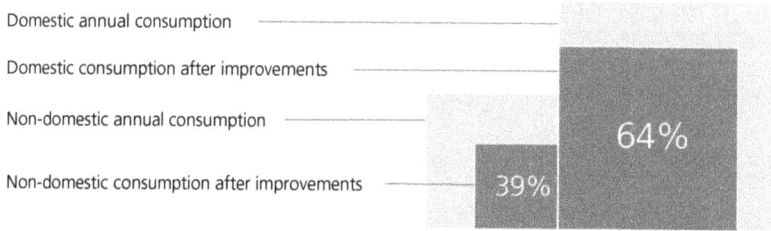

Domestic annual consumption

Domestic consumption after improvements

Non-domestic annual consumption

Non-domestic consumption after improvements

64%

39%

2.3.   The need for flexibility

To improve the way interventions in the built environment are made to ensure they have a positive impact on minimising future emissions and embodied carbon, it is important to explore how they are currently considered prior to their start and, once deemed necessary, how they are briefed to those delivering them. We will cover the topic in more detail in Chapter 11, but it is worth raising here that most construction projects today are designed and delivered with a specific need in mind, which will be met on day one of the asset being used. For example, a new school may be required to educate a specified number of pupils with a known curriculum, potentially with some flexibility to serve a slightly higher or lower cohort. However, where there is a change in the curriculum that the school wishes to deliver, such as the recent introduction of the generally more practical T-levels, there may be a developing need for more heavily serviced or different arrangements of teaching spaces than is typically offered in general teaching spaces. Designing in the flexibility to easily change the level of servicing within a space, or the ability to reconfigure spaces, could have a significant impact on the amount of remodelling, demolition and rebuild required in such circumstances, often at little extra initial effort or cost. Such approaches can certainly make commercial as well as practical sense through the life cycle of an asset and, while not every future scenario can be foreseen or accounted for, briefing or designing out inflexibility can be reasonably simple to achieve.

There are many, many examples of buildings throughout the last century that have been designed with in-use flexibility in mind, from the traditional Japanese house with sliding screens to the Tottenham Hotspur Stadium with its retractable pitch. In infrastructure, consider an airport terminal that suddenly has a significant reduction in demand (for example through the COVID-19 pandemic). The spaces still need to be heated or cooled to ensure the same level of service for users, but the costs may be prohibitive. If a terminal were designed in such a way that say 80% of its space could be closed without reducing the level of service provided, then the asset is designed with flexibility in mind. However, in-use flexibility is different from future adaptability, with the latter requiring some form of remodelling and representing more permanent transformations. A great example can be seen in a large portion of office spaces built over the last 50 years or so which can be reconfigured without changes to the asset's shell. This flexibility is usually enabled by having clearly defined service spaces in floors and ceilings that enable transformations with minimal structural changes. However, when new and potentially unfamiliar building systems are used, it is even more important that future adaptability is considered, as the wrong initial approach can add significant constraints on future adaptability. For example, where some forms of volumetric units are used it can be difficult to remove internal walls in the future as they can be integral parts of a building's structure and service routes.

Figure 2.9 Relative England spend on delivery, operation and services 2021 (Author's own)

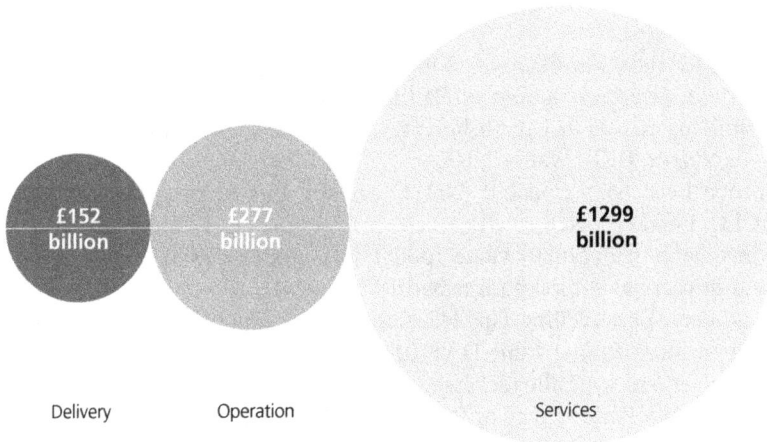

£152 billion

£277 billion

£1299 billion

Delivery   Operation   Services

Figure 2.9 shows the relative money spent on delivering, operating and providing services in built assets in England in 2021. It is clear to see that the amount spent on initial delivery is significantly outweighed by the cost to operate built assets. It leads to the question as to whether services could be provided better, and for less, if built assets kept pace with changing needs, and could be operated more efficiently with more foresight in the initial design and delivery.

If an asset owner were to remain in control of an asset through its life cycle, then it may be quite a simple equation to enable future adaptability at potentially additional cost at the delivery phase to make savings later. However, where ownership of an asset is likely to change then it is clearly harder to justify even a little additional upfront cost if the benefit is unlikely to be realised by future owners alone. This is where the value to society comes in; providing assets which can be upgraded or remodelled to suit changing needs, to reduce future carbon emissions and expenditure, to reduce embodied carbon and enable a more sustainable built environment.

2.4.   Moving forward

In 2015 the United Nations developed its 17 Sustainable Development Goals (SDGs). Of these, almost three quarters are directly or indirectly influenced by the built environment. While the majority of the SDGs are not directly relevant for individual interventions in the built environment, they are at a community level. To provide a clearer link between the SDGs and the decisions made on individual projects, the Construction Innovation Hub (2020) has created its Value Toolkit and the Social Value Portal (2023) have developed the National TOMs Framework. Both aim to support value-based decision making and are described in more detail in Chapter 11.

Most of the first two chapters has been focused on identifying the challenges the construction industry and built environment face and will face in the future. It may seem pessimistic, but the purpose is to be clear on what needs to be addressed to hopefully provide rounded solutions that stand a greater chance of success than if holistic challenges were left unconsidered. As Figure 1.2 illustrates, it is important to clearly understand the issues faced before addressing how they can be overcome. The next chapter looks at the construction industry itself, including recent initiatives that promise to set the industry on a more positive path to delivering a sustainable built environment, then the remainder of the book discusses how automation and manufacturing-led construction can help deliver it.

REFERENCES

BEIS (Department for Business, Energy and Industrial Strategy) (2016) *Building Energy Efficiency Survey, 2014–15: Overarching Report*. BEIS, London, UK.

BRE Group (Building Research Establishment) (2021) *The Cost of Poor Housing in England: 2021 Briefing Paper*. BRE, Watford, UK.

CLC (Construction Leadership Council) (2021) *Greening Our Existing Homes: National Retrofit Strategy*. CLC, London, UK.

Construction Innovation Hub (2020) Value Toolkit. https://constructioninnovationhub.org.uk/our-projects-and-impact/value-toolkit (accessed 18/07/2024).

DLUHC (Department for Levelling Up, Housing and Communities) (2022a) English Housing Survey Data on Stock Profile Table DA1101: Stock Profile. https://www.gov.uk/government/statistical-data-sets/stock-profile (accessed 01/11/2022).

DLUHC (2022b) Energy Performance of Buildings Data: England and Wales. https://epc.opendatacommunities.org/ (accessed 31/10/2022).

HMG (Her Majesty's Government) (2022) Building Safety Act 2022. The Stationery Office, London, UK.

Mayhew L (2020) *Too Little, Too Late? Housing for an Ageing Population*. London, UK.

ONS (Office for National Statistics) (2022) House Price (Existing Dwellings) to Workplace-Based Earnings Ratio. https://www.ons.gov.uk/peoplepopulationandcommunity/housing/datasets/housepriceexistingdwellingstoworkplacebasedearningsratio (accessed 11/11/2022).

Roberts-Hughes R (2011) *The Case for Space: The Size of England's New Homes*. Royal Institute of British Architects, London, UK.

Social Value Portal (2023) National TOMs Framework. https://socialvalueportal.com/solutions/national-toms/ (accessed 16/12/2023).

UKGBC (UK Green Building Council) (2021) *Net Zero Whole Life Carbon Roadmap: A Pathway to Net Zero for the UK Built Environment*. UKGBC, London, UK.

UNEP (United Nations Environment Programme) (2020) *2020 Global Status Report for Buildings And Construction: Towards a Zero-Emissions, Efficient and Resilient Buildings*. UNEP, Nairobi, Kenya.

WHO (World Health Organization) (2011) *Environmental Burden of Disease Associated with Inadequate Housing*. WHO, Geneva, Switzerland.

zu Ermgassen S *et al.* (2022) A home for all within planetary boundaries: pathways for meeting England's housing needs without transgressing national climate and biodiversity goals. *Ecological Economics* **201**: 107562.

Steve Thompson
ISBN 978-1-83608-599-7
https://doi.org/10.1108/978-1-83608-598-020241004

Chapter 3
The construction industry

3.1. What is the construction industry?

While this book primarily looks at the application of automation and manufacturing-led construction on a project or asset level, as with the previous chapters this one begins by looking at the macro level of the industry and its impact on the micro.

The construction industry in the UK makes a significant contribution to the nation. Apart from delivering the built environment that supports mankind, it delivers approximately 6% of gross domestic product (GDP) (£133 bn in 2021) (ONS, 2022a) and directly employs over 2.2 m people (ONS, 2023a). However, these figures do not tell the whole story of the impact or reach of the industry.

For statistical purposes, the construction industry only refers to on-site activities, meaning it misses out key parts of what many see as being essential to the delivery of the built environment. For example, the official statistics do not include professional services such as architectural, structural or civil engineers, nor do they include product manufacturers that only supply into the construction industry, or those responsible for distribution such as builders' merchants. When these are included, the total employed by the wider construction industry increases to around 3.5 m, and it is this wider industry that the book refers to when it talks about the construction industry. This is important, as many of the potential benefits of new technologies or approaches are realised outside of the construction site or project. For example, of over 250 businesses identified within the Government's modern methods of construction (MMC) framework (KOPE, 2023), 55% do not fall into the narrower statistical definition of the construction industry, so the benefits of using such technologies will not appear in the statistics but will impact the delivery of construction projects. The Chartered Institute of Building (CIOB) (Green, 2020) suggests that the wider construction industry is growing faster than the narrower definition of construction and is likely to continue to do so in the future.

Figure 3.1 shows the proportion of people employed in the wider definition of the construction industry, with construction only making up 68% of the total and the remainder being products and services to support construction. The full list of standard industrial classification (SIC) codes that are considered as being part of the wider built environment for the purposes of this book are included in the Appendix.

The value add of the wider construction industry is split along similar lines to the level of employment, with construction product manufacturers delivering 17% of the gross value add (GVA), and 35% of the sector's output. The contractors and subcontractors only deliver 66% of the GVA (Gruneberg and Francis, 2019).

Figure 3.1 Employment in the wider construction industry, 2015 (Source: Gruneberg and Francis, 2019)

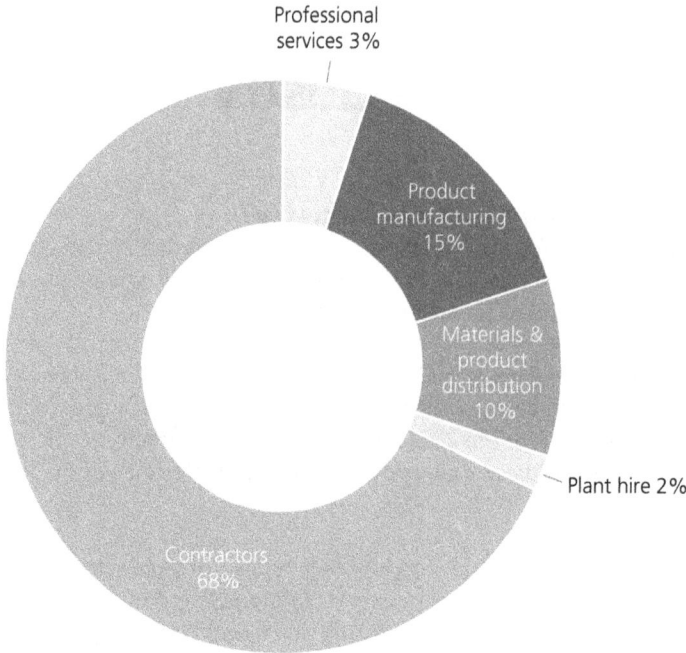

The core of the construction industry is still largely a craft-based industry, where individual projects are designed and engineered to order with low productivity when compared to the economy as a whole, and manufacturing-led industries in particular. This is especially true of the repair and maintenance of existing assets.

3.2. How is the industry structured?

Output from the construction industry can be divided into six main areas. These are infrastructure, residential and other buildings, both new-build and repair and maintenance. Figure 3.2 illustrates the relative value of output from each area in 2021, but it must be noted here that, as has been described in the previous chapters, the industry must change its focus in the coming decades to support the delivery of the Government's net zero targets, with an increased focus on refurbishment of existing assets. That being said, this chapter focuses on describing the construction industry as it is today and the challenges faced, with later chapters identifying how the industry is likely to change in the years ahead.

No industry player, no matter how large, has the knowledge or capacity to provide services across the full depth or breadth of the construction industry. As such, designers, contractors and to some degree product suppliers tend to specialise in particular subsectors. The industry is therefore largely formed of businesses that provide services relevant to specific life cycle stages and specific asset types, with a smaller proportion of businesses that offer multidisciplinary or cross-sector services. In addition to fragmentation by asset type and life cycle stage, the construction market is also largely a localised activity, with projects being location-specific with a requirement for local

Figure 3.2 Construction industry output, Great Britain 2021 (Source: ONS, 2022b)

New build

£46.5bn £29.9bn £39.1bn

Repair and maintenance

£31.3bn £11bn £19.7bn

Residential   Infrastructure   Other

knowledge and capability. As such, many construction businesses are restricted by geography so economies of scale can be difficult to achieve. This level of fragmentation can and does lead to very transactional relationships between businesses on a project-by-project basis, which can lead to low levels of trust. This level of fragmentation, and the culture that it develops, can provide a real challenge when it comes to transforming the industry, and it is therefore likely that the pace of transformational change will vary widely across market segments and potentially geographies.

While the supply chain for each market segment or geography can be different, it is broadly accepted that most construction supply chains are very complex and fragmented. However, at its essence, there are five key elements that are required to deliver a new product, whether that product be a new brick wall or a completed asset. The supply chains involved in the delivery of construction can then be described according to these five elements, which are identified in Figure 3.3.

Where the supply chain becomes complex is the sheer number of products and services that are delivered by different parties on any given construction project, and when these products and

Figure 3.3 The Five Ps model (Author's own)

Process – the activity being carried out to deliver the new product, e.g. laying bricks to build a wall or constructing a whole new building

Plan – the professional services required to design, plan, procure and manage the activity

Product – the physical products and materials required as input to the activity

People – the human resources required to carry out the activity, e.g. to build the wall or to construct the building

Plant – the tools required to carry out the activity, e.g. a screwdriver or a tower crane

services are aggregated to provide the finished asset. This method of looking at supply chains becomes particularly relevant when looking at the potential impact of new technologies or ways of working. As a simple example, a particular block may be more expensive than a rival product but may be quicker to install, with less resource and less manual lifting required, so the desired output is potentially delivered quicker, cheaper and safer.

In construction, more so than in most industries, the end product (the completed asset) tends to be unique, with the result of every project being different in some way. On top of that, the team assembled to deliver each project is usually formed for that project specifically, so the structure created only exists for a short period of time before it is dismantled, with a new team created for future projects. This means that the points at which products and activities are aggregated changes from project to project, so there is no consistency in the roles played by different parties, or what they are delivering to whom.

In 2014, the Department for Business, Innovation and Skills (BIS, 2013) analysed the construction supply chain and identified that it was not unusual for Tier 1 organisations in projects to have in excess of 70 subcontracts, and for each of those subcontracts to have 40 subcontracts themselves. Over 50% of intermediate spend was on subcontracting, which is the highest of any industry. The value of contracts between Tier 1 main contractors and Tier 2 suppliers analysed varied significantly, however, with typically a handful of large Tier 2 subcontractors delivering over 75% of the value of a project. The remaining 25% was then delivered by a long tail of much smaller contracts, depending on the specialism required. However, most large Tier 2 suppliers subcontracted the work to a disaggregated Tier 3 supply chain, where most of the work was carried out. That suggests that it is not uncommon to have two layers of management and procurement between the client and the team carrying out the work – in other words, referring to Figures 3.3 and 3.4, two layers only

Figure 3.4 Analysing construction supply chains (Author's own)

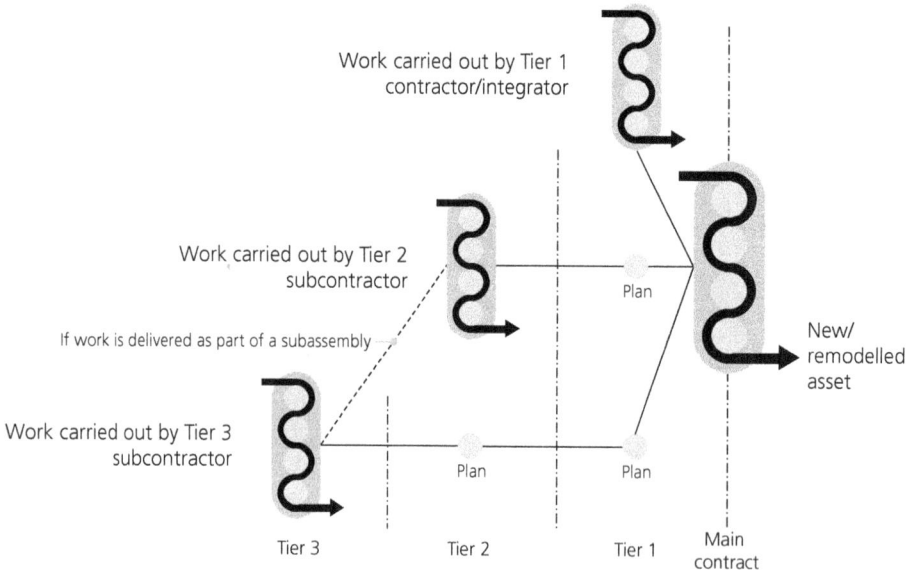

focusing on the 'plan' element of the process, and not on the other four key value add elements, meaning that the delivery of the process is inefficient.

Where feasible, an activity should have as few layers of direct management as possible unless it is integrated into a subsystem, as shown in Figure 3.4. It must be emphasised here, however, that Figure 3.4 illustrates the process of delivery of an asset, not contractual relationships. There are valid reasons to subcontract the management or coordination of work – for example, to manage risk – but that does not mean that there are not more efficient ways to deliver projects.

Mapping out supply chains in this fashion can be very useful in identifying where aggregation should occur – for example, where systems should be utilised and potentially where contractual boundaries make most sense. These mapping processes can be undertaken digitally and will be discussed in later chapters. Some of this may involve understanding where systems are made-to-order, against engineered-to-order (i.e. delivering standardised against unique products).

Designers and professional consultants tend to be appointed in isolation of the construction team, who are traditionally engaged late in the development process, and with a focus on bidding based on lowest capital cost and pushing risk down the supply chain. These issues are not necessarily resolved where a supply chain is partially integrated (for example, by incorporating an off-site manufacturer), as fragmentation can remain in the rest of the supply chain if it is not fully considered. Combined with the fact that, traditionally, facilities management teams are not typically involved in the development or appraisal process until projects near completion (unless a soft landings[1] approach is used), this also leads to fragmentation along the duration of an asset's life cycle. As such, it can be challenging to develop fully integrated designs that consider the whole life cycle of an asset or maximise value delivered to the client.

The level of fragmentation within the supply chain leads directly to increased costs, due to the cost of businesses transacting with each other. The Institution of Civil Engineers (ICE, 2017) suggests that for infrastructure projects the overhead and profit for Tier 1 contractors is around 5%, with additional management costs of 16% covering project management, design and commercial. For Tiers 2 and 3, the overheads and profit can be as much as 13.5%, with an additional 7–14% for management. The additional management for lower tiers is understandable, as they are more likely to employ direct skilled labour and directly own plant. Gruneberg and Francis (2019) suggest that profit margins for main contractors are in the order of 1.5%, but as high as 15% for specialist contractors. While figures clearly vary from project to project, these present a clear picture of an industry working on low margins at the main contractor level in particular. Instead, the large contractors make their money by carefully managing cash flow, through disputes and additional work, and through working on multiple projects simultaneously. However, these models do add fuel to the fire of an already litigious industry. Still, despite the potentially low profit margins, the amount the owner client spends on management, overheads and transaction costs can be more than 50% of the total cost of a project (ICE, 2017), a figure corroborated by Bryden Wood (2021). So, while there is certainly good reason for some of this spend, there are surely efficiencies to be gained.

---

[1] 'Soft landings' is a methodology to ensure that the future operation of an asset is considered through the development process. This has developed further into the Government Soft Landings (Cabinet Office, 2013).

The supply chain analysis by BIS (2013) suggested that the main opportunities to reduce costs and increase efficiency in the supply chain were likely to be

- early contractor and subcontractor involvement in solution development
- coordination of design and assembly across the supply chain, potentially based in sharable asset information – for example, BIM
- effective management of change
- effective on-site operations, supported by an integrated, established site team
- wider adoption of the 'integrator' role of supply chain management, either at Tiers 1 or 2.

While these are all laudable initiatives, the industry's move to a more value-driven, outcomes-based model by using approaches such as the Construction Innovation Hub's (2023) Value Toolkit, are also very valuable in ensuring that the right assets are delivered in the right way. Without that holistic, unified and consistent focus on delivering a common goal, it can be challenging to deliver the best outcomes and can lead to inefficiencies. Traditionally, the main contractor provides the overall planning for a project, but their subcontractors develop their own detailed plans and logistics. This can mean that no one organisation has a fully detailed view of what is happening and the lack of a common plan can lead to inefficiencies, with defects and errors adding up to 20% of the cost of a project in some cases (ICE, 2017).

Another sign of the fragmentation of the construction industry is the number and size of businesses it contains. Over 96% of businesses within construction employ fewer than ten people, driven largely by a high number of self-employed contractors. However, the remaining 4% deliver 83% of the output (ONS, 2022b). This figure does vary by subsector, however, as shown in Figure 3.5, with smaller businesses delivering most of the work in housing repair and maintenance.

The average size of business also changes when the wider construction industry is considered, as shown in Figure 3.6. In construction product manufacturing for example, only 76% of businesses have fewer than ten employees, and 85% of merchant businesses (ONS, 2023b).

Figure 3.5 Great Britain construction output (£million) by company size (Source: ONS, 2022b)

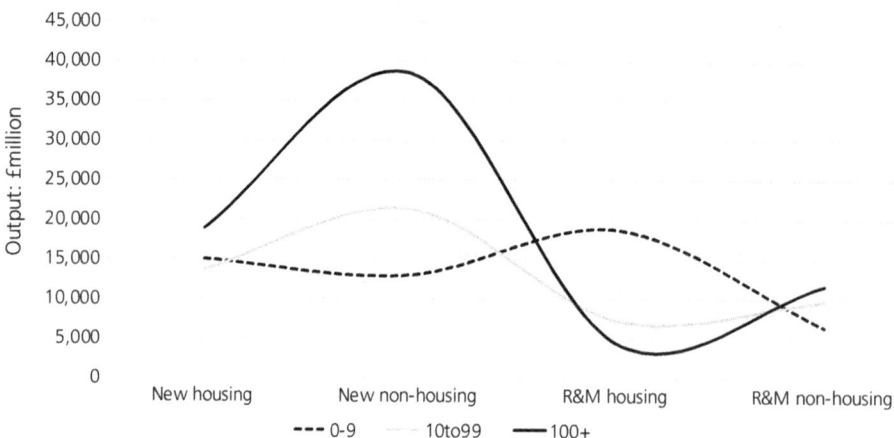

Figure 3.6 Business size and turnover by subsector, Great Britain, 2021 (Source: ONS, 2022b)

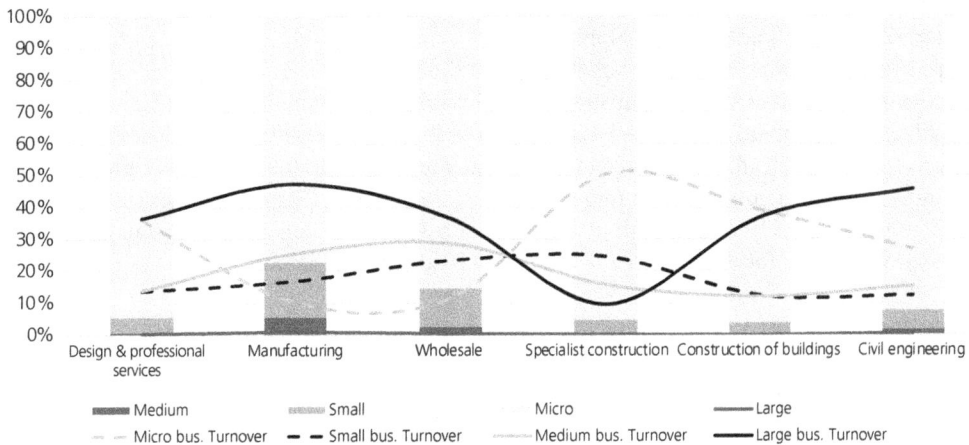

These figures suggest that the current integration of activities on a construction site is more likely to be delivered by smaller businesses, meaning that larger businesses, such as product manufacturers and main contractors, rely heavily on smaller specialist businesses to deliver their value. This means that no matter what the size of a manufacturer, it needs to have supply routes that enable its products to be delivered in small quantities to individual projects, while still achieving economies of scale in production. It also means that it may be harder to achieve economies of scale for those delivering projects, as smaller businesses tend to work on one or two projects at a time, so do not have the scale to receive or pass on that benefit. In addition, research shows that smaller businesses are less likely or slower to adopt digital technologies on average as they do not necessarily have the resources to invest in research and development, or the spare resources to focus on strategic development instead of project delivery.

Referring to Figure 3.2, the widespread delivery of repair and maintenance work by smaller businesses illustrates part of the challenge in adopting new ways of working on existing assets, combined with the complications of dealing with unique assets against delivering new. It isn't that small businesses cannot or do not want to adopt new technologies, but that they do not always have the opportunity to do so. Part of the challenge that needs to be addressed to move the industry forward then is to support small- and medium-sized enterprises (SMEs) in adopting new technologies and ways of working. There are many SMEs who are proactive in their use of new technology, but they need to have the right partners or clients to flourish and take advantage of its potential. With the traditional business model in construction, it is normally the case that subcontractors are appointed based on lowest cost to deliver a service, not necessarily on the added value that they can deliver.

3.3. How does the industry perform?

So, given all the structural challenges raised so far, how does the industry currently perform? Well, there have been many well-publicised cost and time overruns on major projects in recent years, including Crossrail in the UK. On average, in 2019–20 only 68% of projects came in on or within budget, with the predictability of costs only at 61% (Glenigan, 2021). That suggests significant challenges in delivering within budget, which is corroborated by Aljohani *et al.* (2017), who found that of 258 transport projects across 20 countries, nine out of ten projects went over budget, with an average of a 28% overrun.

Glenigan (2021) also found that programming was a challenge, with only 60% of projects being delivered within programme. However, at the same time, over 90% of clients rated their finished product as 8 out of 10 or better. That suggests that, while assets may often be delivered late and over budget, the quality of the finished asset is high. But imagine if only 60% of your online deliveries arrived on time, or if, when you got to the till in a supermarket, the cost of your groceries was nearly 40% more than you had been told they were going to cost. Clearly the construction supply chain is more complex, but such high percentages of assets being delivered over time and budget does cause real issues for clients and end users, and is something that needs to be addressed.

The construction industry also struggles with late payment, especially SMEs who often rely on prompt payment to manage their cashflow. Glenigan (2021) found that only 71% of projects valued at less than £1 m rated payment as being 8 out of 10 or better. It is not uncommon then, for small businesses to rely on credit from builders' merchants to help them manage these issues.

Bearing in mind the narrow definition of construction used in official statistics, Figure 3.7 shows the productivity of different sectors within the industry over the last 25 years.

Apart from specialist construction, productivity has fallen in the last two decades, and the gap between construction and the economy has increased. The largest falls in productivity have been in the construction of buildings and design activities, with specialist construction rebounding to be slightly above where it was at the end of last century.

Figure 3.7 Construction productivity growth comparison (UK), output per hour worked (Source: ONS, 2022c)

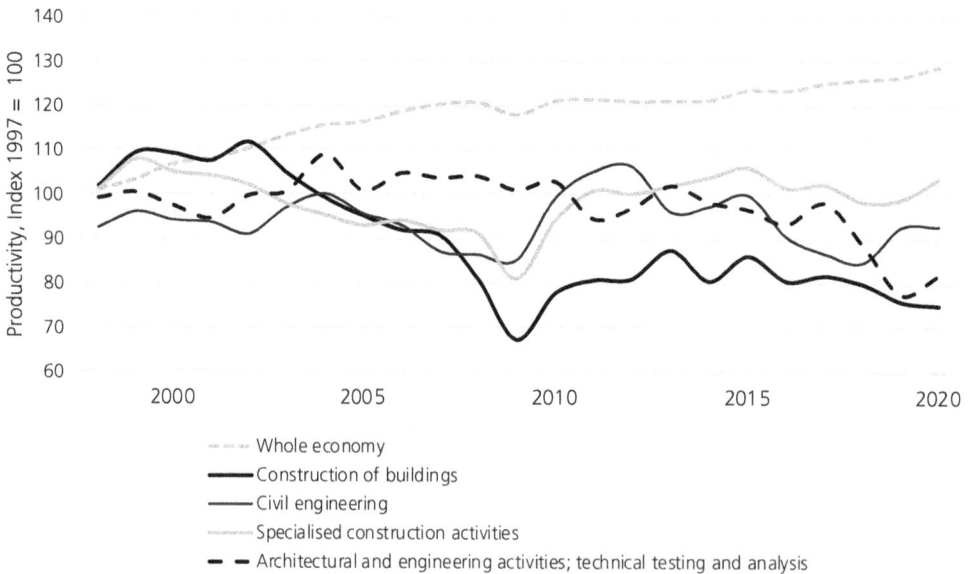

There is often a lot of discussion suggesting a move to a more industrialised approach to construction will help address the low productivity within the industry, and this book will look to test that logic in later chapters to see how that can become a reality. As a starting point, however, it must be recognised that the construction industry today is still largely craft-based, not manufacturing-based, so the potential shift is a significant one.

In looking to address the potential shift, it is important to look at some of the challenges faced, here using the elements identified in Figure 3.3 to describe them.

3.4.   Process

Starting with process, much of the construction industry still builds assets predominantly the way they were built decades ago. This is particularly true in the residential sector, and in the remodelling of existing homes even more so. To some degree that is driven by demand, in that homeowners often have a preference for a traditional aesthetic using traditional materials, whereas in other, more consumer-oriented industries, products have adapted more to suit the latest materials and technologies. However, it is also driven by other factors, such as the limited ability to innovate on individual projects and to take that learning onto future projects with different teams, with potentially limited incentive to change. If the change can be made, the potential opportunities are significant. For example, for every 1% increase in productivity, a 1.5% increase in real wage growth can be achieved (Mace, 2021).

A further consideration is whether the right processes are being used to deliver the required outcomes. An example of this is construction flexibility, so are the processes optimised for delivery under the given circumstances, or are they selected based on other criteria, such as availability or familiarity with a particular contractor or consultant? For example, are volumetric modules being used where a panel system would suit the requirements better because of site access or the need to be able to make late changes to finishes or fixtures due to client requirements?

3.5.   Plan

Change is now a constant within the construction industry. While industrialisation of other industries happened gradually over decades, the construction industry is expected to transform at pace, largely because of the many challenges it faces. There are global challenges in terms of competition and supply chain issues brought about by global conflicts or extreme weather events, and changing demand in terms of what clients are looking for (changes in asset requirements, lower carbon, quicker and cheaper assets). Thankfully, there is also rapid technology development, but the technology is often developing much quicker than the industry can cope with, so business models and the pace of skills development are not necessarily able to cope with the change. This is where planning comes in. Not planning just for known changes, but also adapting the way change is managed.

The old change management processes of having a clear blueprint of a future state, a picture of the current state and a clear plan on how to get from one to the other is not likely to be sufficient to cope with the pace and level of change that is likely to occur in the coming decade. This is where lessons can be learned from the technology sectors, where an agile approach to managing change is used and developments are constantly reviewed to ensure they meet changing requirements.

The planning approval process in the UK represents a significant challenge in enabling change to a more efficient construction industry, and not just because of the time it takes to grant permissions or the uncertainty that it causes. As described in later chapters, a key change the industry needs to make is to move towards considering construction projects as part of a portfolio of interventions that support the achievement of defined outcomes. These broader considerations are likely to include looking at the value that can be delivered over a number of interventions, not just over a single project, but the current planning process only considers projects on an individual basis.

Another regulatory change that will have a significant impact on the construction industry is the Building Safety Act 2022 (HMG, 2022), developed in response to the Grenfell Tower tragedy of 2016. The Act is likely to have far reaching implications for the industry, through additional regulation of construction products and practices. While the changes will certainly increase costs and timescales in some areas (and the potential for an increase in disputes), they also provide another push towards better information management and digital practices and help to ensure better outcomes for asset owners and users alike. The Act is discussed in more detail in Chapter 10.

Finally, there is a real need to work with clients to ensure that they are clear on what it is that they really need. As Theodore Levitt once said, 'people don't want to buy a quarter-inch drill, they want a quarter-inch hole' (Christensen and Raynor, 2003). It is important to work with clients to clearly define their short- and long-term needs and to make clear the benefits of long-term solutions, not just the delivery of the lowest initial construction cost. As an example, it is unlikely that most homeowners of newly built homes are aware that they will need to undertake expensive refurbishments within the next two or three decades if their homes are to achieve the Government's net zero requirements. It can be a real challenge convincing those with a short-term interest to provide more sustainable solutions, such as speculative developers who are unlikely to benefit from any increased expenditure in providing solutions that only add value over the long term. There is a big question on who pays to achieve a net zero built environment; it is likely that by the late 2030s the savings in operating costs will outweigh the costs in getting there if investment is made to achieve it, but who finances the significant upfront cost? Part of the challenge, then, is delivering whole life solutions while minimising initial construction costs, and later chapters will discuss how manufacturing-led construction and careful planning can support this.

3.6.　People

The skills challenges faced by the construction industry are well documented, with nearly 45 000 new workers forecast to be required per year over the next five years to meet expected output (CITB, 2021). As well as meaning that there are potentially not enough workers to carry out planned construction, it can lead to premiums being paid for the skills in demand and a knock-on effect of increased construction costs. One reason for the shortage is the ageing of the current workforce, with over 31% of the current workforce aged over 50 years. The two main reasons this is an issue are that many are likely to retire in the next decade or so, but also the conditions that existing construction projects are delivered in are often not suitable environments for older individuals to operate in. However, as Figure 3.8 indicates, there is a higher percentage of older workers in the construction manufacturing sector, and that is partly because of the nature of the environment; an indoor, controlled environment meaning that workers can stay employed for longer.

Another challenge in addressing the skills shortage is attracting new people to the industry. With the ever-increasing use of technology, the industry is more frequently competing with other industries for talent, with over a million job vacancies in 2022. It can be a challenge to attract new workers when the industry is still seen by many as being a low-skilled or low-paid career, often carried out in dirty and dangerous environments, and one that is project-based, so may not have long-term

Figure 3.8 Age profile of construction workers (UK), 2021 (Source: ONS, 2021)

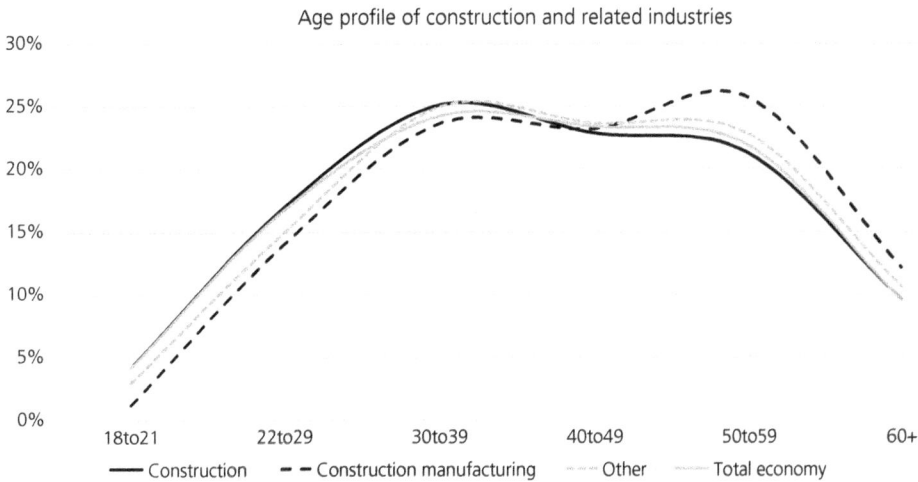

Age profile of construction and related industries

Construction — Construction manufacturing — Other — Total economy

job security. Thankfully, as the industry moves to a more industrialised model, the potential for better working conditions and high-quality jobs is increased, with Mace suggesting that more than 124 000 jobs can be created through setting up regional manufacturing hubs (Mace, 2021).

It is important to recognise that, with the application of new technologies and manufacturing-led approaches, new types of jobs will be developed which will require new skills or combinations of existing skills, blurring the lines between existing roles. With the rapid development of technology, these changing skills requirements are likely to develop in the next five years in some cases, putting pressure on the existing workforce, training providers and employers to adapt accordingly. The impact of automation and manufacturing-led approaches will be discussed at length through this book but is likely to be a case of augmenting existing skills and repackaging activities into new roles, instead of replacing jobs within the industry.

The culture of the construction industry is one of the most significant challenges faced when bringing about change. The disruption necessary to take the industry forward is not likely to require a gradual, incremental shift away from existing practices and models, but quick (within two or three years) reframing of existing business models. Within an industry that has not changed significantly for such a long time (and has survived under the existing model), it can be hard to accept that the change is necessary or beneficial, and the temptation can be to delay it. However, with the challenges now facing the industry, many described in this book, the industry is likely to change. If not with the existing players, then potentially with new entrants who seize the opportunities that exist.

3.7. Product

The physical built environment is made up of combinations of construction products and as such, material and product manufacturing and supply are crucial parts of the industry. The early 2020s have seen very significant material shortages and subsequent price rises due to several global issues, which have had a knock-on effect on the cost of construction. Figure 3.9 highlights the relative speed and level of change for some products, which have caused many projects to stall as they

Figure 3.9 Construction material price indices, UK (Source: BEIS, 2022)

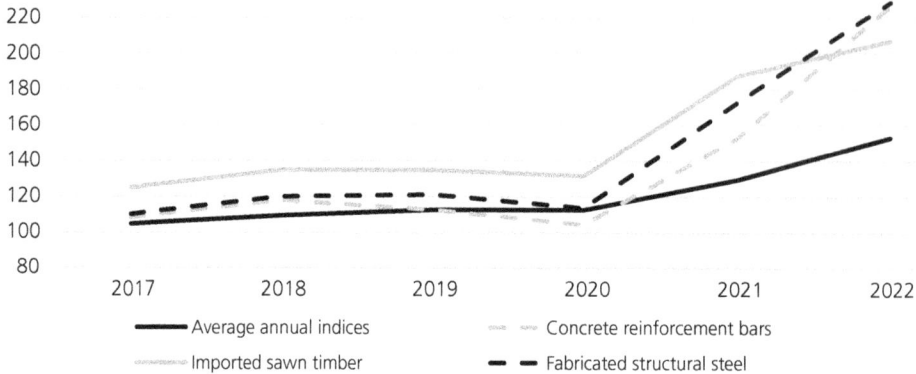

Legend:
— Average annual indices
··· Concrete reinforcement bars
··· Imported sawn timber
– – Fabricated structural steel

have become unviable, and many businesses (subcontractors in particular) to fall into administration as they have delivered fixed-price projects agreed before the rise.

These material shortages have also led to hoarding of stock in some cases, meaning that to avoid lengthy delays and increased risk, early-stage planning is required with the supply chain for new projects. The traditional business model for construction tends to prevent owners and consultants from engaging directly with suppliers, but it is ever more important to do so when shortages occur. While 75% of the construction products used in the UK are made in the UK (CPA, 2023), the trade gap between imports and exports has been increasing over the last decade. While Brexit is expected to have an impact on this, many products and raw materials will still be more economical to make elsewhere and import, meaning that they will still be impacted by global supply chains and continue to face challenges.

3.8. Plant

The plant hire sector turns over nearly £4 bn and employs over 43 000 people in the UK (Gruneberg and Francis, 2019). With such low profit margins, Tier 1 contractors typically do not own the plant they use, instead relying on their subcontractors to either use their own or hire the necessary plant. This means that the plant hire sector is highly fragmented and, as with other parts of the industry, works on a project-by-project basis, meaning that it is not necessarily the most efficient way to procure plant. With individual subcontractors hiring plant, it can be difficult to optimise utilisation of equipment across a project, never mind several projects at once. However, there are examples of plant being procured at the project level and shared collaboratively as required for the good of the project, which can lead to significant savings.

As technologies develop, it is increasingly important to have access to high-speed internet, but the UK has poor access when compared with other countries. However, the National Infrastructure Strategy (HMT, 2020) is looking to address this by supporting UK-wide gigabit broadband roll-out. Digital technologies and automation promise to make a significant contribution to the move to net zero, but to enable this to happen, the infrastructure needs to be in place.

3.9. Moving forward

Over the last few years, green shoots of change have been developing. The Government's BIM strategy has been implemented and the focus moved on to better information management. The

Construction Innovation Hub, supported by the Government, have developed an approach to product platforms for the construction sector, as well as the Value Toolkit, which supports clients in defining their desired values (Construction Innovation Hub, 2023). The Government's Construction Playbook (HMG, 2020) has also set out future delivery requirements for public projects, again focused on delivering better outcomes. So, things are moving in the direction of a more outcomes-focused, systems approach to delivery and maintenance of the built environment, with a focus on the whole life of an asset, not just its initial construction. With p-DFMA (platform design for manufacture and assembly), there is also movement towards manufacturing-led construction, supported by the 'presumption in favour of off-site' applied to public projects. The biggest change required in moving towards these new approaches is cultural, as many of the necessary technologies already exist, albeit not necessarily in abundance. A shift to more collaborative working has been proposed for several decades as a way of improving the performance of the industry, but with the significant challenges the industry now faces, and with potential solutions within touching distance, the concept is more likely to become reality.

With the UK being recognised as a more expensive place to deliver infrastructure projects (HMT, 2010), the shift to a more productive model needs to happen for the UK to remain competitive in other industries, not just in construction. The remainder of the book will now look at potential solutions to the challenges and issues raised so far.

REFERENCES

Aljohani A, Ahiaga-Dagbui M and Moore D (2017) Construction projects cost overrun: What does the literature tell us? *International Journal of Innovation, Management and Technology* **8(2)**: 137–143.

BEIS (Department of Business, Energy and Industrial Strategy) (2022) *Monthly Bulletin of Building Materials and Components*. BEIS, London, UK, May.

BIS (Department for Business Innovation and Skills) (2013) *Supply Chain Analysis into the Construction Industry*. BIS, London, UK.

Bryden Wood (2021) *Delivery Platforms for Government Assets*. Bryden Wood, London, UK.

Cabinet Office (2013) *Government Soft Landings – Executive Summary*. Cabinet Office, London, UK.

Christensen CM and Raynor ME (2003) *The Innovator's Solution: Creating and Sustaining Successful Growth*. Harvard Business School Press, Boston, MA, USA.

CITB (Construction Industry Training Board) (2021) *Construction Skills Network 5-year Outlook 2021–2025*. CITB, London, UK.

Construction Innovation Hub (2023) The Value Toolkit. https://constructioninnovationhub.org.uk/our-projects-and-impact/value-toolkit/ (accessed 16/12/2023).

CPA (Construction Products Association) (2023) https://www.constructionproducts.org.uk/ (accessed 16/03/2023).

Glenigan (2021) *2019/20 UK Industry Performance Report*. Glenigan, London, UK.

Green B (2020) *The Real Face of Construction 2020*. Chartered Institute of Building (CIOB), London, UK.

Gruneberg S and Francis N (2019) *The Economics of Construction*. Agenda Publishing, Newcastle upon Tyne, UK.

HMG (Her Majesty's Government) (2020) *The Construction Playbook*. The Stationery Office, London, UK.

HMG (2022) Building Safety Act 2022. The Stationery Office, London, UK.

HMT (HM Treasury) (2010) *Infrastructure Cost Review: Main Report*. HM Treasury, London, UK.

HMT (2020) *National Infrastructure Strategy: Fairer, Faster, Greener*. HM Treasury, London, UK.

ICE (Institution of Civil Engineers) (2017) *From Transactions to Enterprises*. ICE, London, UK.

KOPE (2023) MMC Market. https://mmc.market/suppliers (accessed 16/12/2023).

Mace (2021) *The New Normal*. Mace, London, UK.

ONS (Office for National Statistics) (2021) *Annual Survey of Hours and Earnings (ASHE)*. ONS, London, UK.

ONS (2022a) GDP output approach, low level aggregates, UK. Quarter 1 (Jan to Mar) 2022, ONS, London, UK.

ONS (2022b) *Construction Statistics Great Britain 2021*. ONS, London, UK.

ONS (2022c) *Productivity in the Construction Industry, UK: 2021*. ONS, London, UK.

ONS (2023a) *EMP13: All in Employment by Industry*. ONS, London, UK.

ONS (2023b) *Specified divisions by employee size*. ONS, London, UK. https://www.ons.gov.uk/businessindustryandtrade/business/activitysizeandlocation/adhocs/15656specifieddivisionsbyemployeesize (accessed 18/07/2024).

Steve Thompson
ISBN 978-1-83608-599-7
https://doi.org/10.1108/978-1-83608-598-020241005
Emerald Publishing Limited: All rights reserved

Chapter 4

Why automation and manufacturing-led construction?

4.1. Introduction

Before diving into why automation and manufacturing-led construction provide potential solutions to the many challenges that face the built environment, it is important to define what is meant by those terms and others that will be used throughout this book. There are very important distinctions to be made between terms that are often used interchangeably for the impact of different technologies and approaches to be clearly identified. Figure 4.1 illustrates the relationship between the following terms

- *Construction* – the core construction industry as defined by the Standard Industrial Classification (SIC) for statistical purposes. This may include the use of modern methods of construction (MMC) where it does not materially change how an asset is delivered and is used on one project in isolation. It can also include the use of design for manufacture and assembly (DFMA).
- *Built environment sectors* – the wider built environment including construction, manufacturing, MMC providers, professional services and asset management.
- *Manufacturing-led construction* – construction where manufactured systems or manufacturing processes are used and materially change the way an asset is delivered from a traditional on-site model, so includes MMC, off-site construction and DFMA. Manufacturing-led construction does not necessarily mean the whole asset is considered as a system, so may include the use of technologies for only a part of the process, with traditional approaches for the remaining activities. It includes one-off projects as well as repeatable products and processes across multiple projects and includes the use of product platforms.
- *Automation* – digital or physical technologies are used to remove the need for human involvement. This does not mean replacing humans but augmenting human activities and realigning activity.
- *Industrialised construction* – the combined use of building assemblies and automation technologies to deliver virtually or totally complete assets. The construction process is considered from end to end, not just in silos. It includes product platforms where they are used for most of an asset, and the use of repeatable products and processes across multiple projects. This is more like manufacturing an asset than delivering manufactured systems into a construction project.
- *Industrialised built environment* – the use of physical and digital systems to deliver, maintain and reconfigure or upgrade assets throughout their life cycle, with very little to no abortive work in a plug-and-play fashion. For example, replacing envelope elements to upgrade the thermal performance of an asset or relocating internal walls to reconfigure internal layouts without the need for additional building work. An industrialised built environment may include the provision of a built environment as a service, whereas industrialised construction is only the delivery of an asset.

Figure 4.1 Relationships between industries, automation and manufacturing-led construction (Author's own)

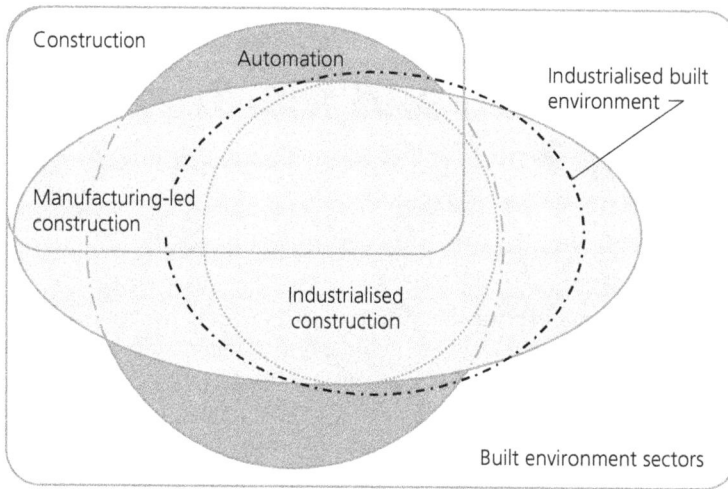

It is important to emphasise the difference between the use of off-site construction, manufacturing-led construction and industrialised construction. Not all uses of MMC fall into manufacturing-led construction for example, as not all applications of such solutions materially affect the way an asset is delivered. For example, the use of a proprietary internal partitioning system may be used on a construction project, but that does not usually significantly alter the way that other aspects of the asset are delivered. However, if prefabricated bridge segments are made off-site, or a volumetric system is used to deliver rooms within a hotel, it materially impacts the way the rest of the asset is delivered. If that hotel were delivered also using manufactured systems for the nonmodular aspects, and the solutions were configurations of technologies common to multiple projects, then it would fall into industrialised construction.

As Figure 4.1 suggests, there is significant overlap and interdependency between automation and manufacturing-led construction, but in considering how they work together and why they should be considered as part of the solution, it is worth first looking at them separately. As identified in Figure 1.2, before looking at automating something, it is important to first ensure that the right thing is being automated, so manufacturing-led construction is the starting point.

4.2. Manufacturing-led construction

In 2017 the UK Government as the UK's largest construction client stated that on all public projects going forward there would be a 'presumption in favour' of off-site construction (HMT, 2017). The commitment to off-site has continued in both the *Transforming Infrastructure Performance Roadmap* (IPA, 2021) and the *Construction Playbook* (HMG, 2020) and so manufacturing-led construction, at least in some form, will be a key part of the delivery of Government assets in the years to come. It is important, however, that the needs and specifics of any intervention in the built environment is carefully considered before being delivered if it is to be successful, whether using manufacturing-led construction or not. As the Institution of Civil Engineers identified in their review *Systems Approach to Infrastructure Delivery* (SAID) (ICE, 2020), projects today should be viewed as interventions to deliver desired outcomes, not just physical assets alone, as has often been the case in the past. Projects today are just as much about the technologies needed for operations as they are about the delivery of physical assets – for example, a railway requires digital signalling systems, timetabling and health and safety communications. As such, projects can be seen as systems of systems, of which the physical asset is an important part. Therefore, delivering systems of systems requires new approaches to those used to deliver a built asset alone. For example, the Channel Tunnel project in the 1980s changed its focus from delivering a tunnel to delivering a railway, two vastly different challenges.

Later chapters will look at the implications of, and methodologies used in, operating a systems approach to delivery, but manufacturing-led construction involves a change in mindset to more holistic definitions of requirements and delivery. Once those requirements are clearly defined, then manufacturing-led construction promises significant opportunities to improve on the performance of the construction industry. According to Mace (2021), over £2.8 bn could be saved across government assets if they were delivered using MMC, while also creating 125 000 regional manufacturing jobs. Mace also suggests that manufacturing-led construction could reduce the cost of infrastructure delivery by 17%, while McKinsey (2019) believes that it could deliver potential savings of $22 bn across Europe and the USA. McKinsey also suggests that, when combined with automation and digitalisation, cost savings can exceed 30%. A real-world example is the A453 road widening scheme delivered by Laing O'Rourke and WYB, where a six-month time saving and 30% reduction of on-site labour was achieved by using DFMA and manufacturing bridge components off-site for assembly on-site by smaller teams than would traditionally be the case. Manufacturing-led construction can not only offer potentially significant cost and time savings, but can also provide greater certainty on both fronts, with less potential for delays due to external factors such as weather.

Another significant, and potentially the main selling point for the use of manufacturing-led construction, is the quality of the end product – the built asset. With the increased precision of production in a controlled environment and the resulting improvement of tolerances, building systems can deliver improved structural and thermal performance, for example. Octopus Energy now work with providers to deliver zero bill homes, where the performance of the building envelope is pivotal to being able to provide that level of guarantee. At the same time, efficient production can significantly reduce waste and reduce embodied carbon. A recent study by the University of Cambridge and Edinburgh Napier University (Inside Housing, 2022) showed that embodied carbon savings of 45% had been delivered on two high-rise modular developments and, while carbon savings clearly vary by form of construction, the efficient use of materials and quality of construction achieved under factory conditions far exceeds that of a traditional construction site. This potential improvement in asset performance combines very well with the recent move towards considering the whole life value of assets, as the increased quality has long-term benefits way beyond the construction phase, including improved energy efficiency and repeatability. However, defects can still occur with prefabricated solutions, most likely where the interfaces between such systems and more traditional construction activities has not been effectively managed.

Manufacturing-led construction also offers significant value in terms of productivity during the delivery phase of a project. Make UK (2022) suggest that modular construction can be over 40% more productive than traditional construction in terms of hours worked per metre squared of floor area, while Mace (2021) believe improvements could be as high as 50%, with each 1% increase in productivity leading to a 1.5% increase in real wage growth.

So, with all these potential upsides, you may ask why manufacturing-led construction hasn't already largely replaced the traditional construction model. There are several reasons why, but with the advancement of technologies (both digital and physical) over recent years and the shift towards whole life value and support from Government, the use of such approaches is certainly becoming more pervasive and will continue to do so.

One major reason why manufacturing-led construction has not become widespread sooner is poor visibility of demand needed to keep expensive production lines running efficiently and, with most projects traditionally being seen as unique, the volume required to sustain factories can be difficult to realise. As the move towards more manufacturing-led construction has developed over recent years, there have been some highly publicised business closures, such as Urban Splash House and Ilke Homes in the UK and Katerra in the USA. However, at the same time, other businesses have seen growth. The Government's published construction pipeline, the promotion of product platforms, the advancement of technologies enabling micro factories, and asset-light approaches to delivery are all potential solutions to alleviate this demand problem and will be covered in Chapter 14.

Another issue relates to interfaces between manufacturing-led suppliers and traditional construction players. First of all, there are physical interfaces to deal with; the precision engineered, right-first-time manufactured solutions produced in a factory meeting craft-based trades used to following-on from other traditional trades. There are also understandable feelings of being threatened by solutions that may take work away from traditional businesses or trades and concerns that manufacturing businesses may remove the need for intermediaries between themselves and clients. The truth, however, is that there are many different manufacturing-led models, each requiring some form of integrator or assembler and, as with automation, manufacturing-led construction will not replace existing trades or contractors completely any time soon but will augment the skills that they have to enable a more efficient use of resources. The interface issues raised can all be dealt with through careful consideration of the delivery process from end to end.

A further challenge is that of self-interest; why would an incumbent want to change to a new model if they already have a successful business and are comfortable with current delivery practices? The answer here is that clients are becoming more intelligent and have access to more information than ever before, meaning the potential for alternative solutions is becoming ever more visible. The results of a McKinsey survey in 2020 found that over 60% of executive respondents believed that a shift to more productive, manufacturing-led approaches would occur at scale before 2025 (McKinsey, 2020). These changes mean that unless incumbents re-evaluate their current business models and adapt, they face the very real risk of losing out on their share of the market in the near future.

Figure 4.2 illustrates some of the potential changes in relationships between industry players as projects move from traditional models to manufacturing-led, through to industrialised construction more specifically.

In the traditional model, the only relationships that the client typically has are with the design team initially, and the main contractor later in the process. The specialist contractors who usually undertake most of the work never interface with the client and are not usually aware of

Figure 4.2 Changing relationships and points of interaction by construction type (Author's own)

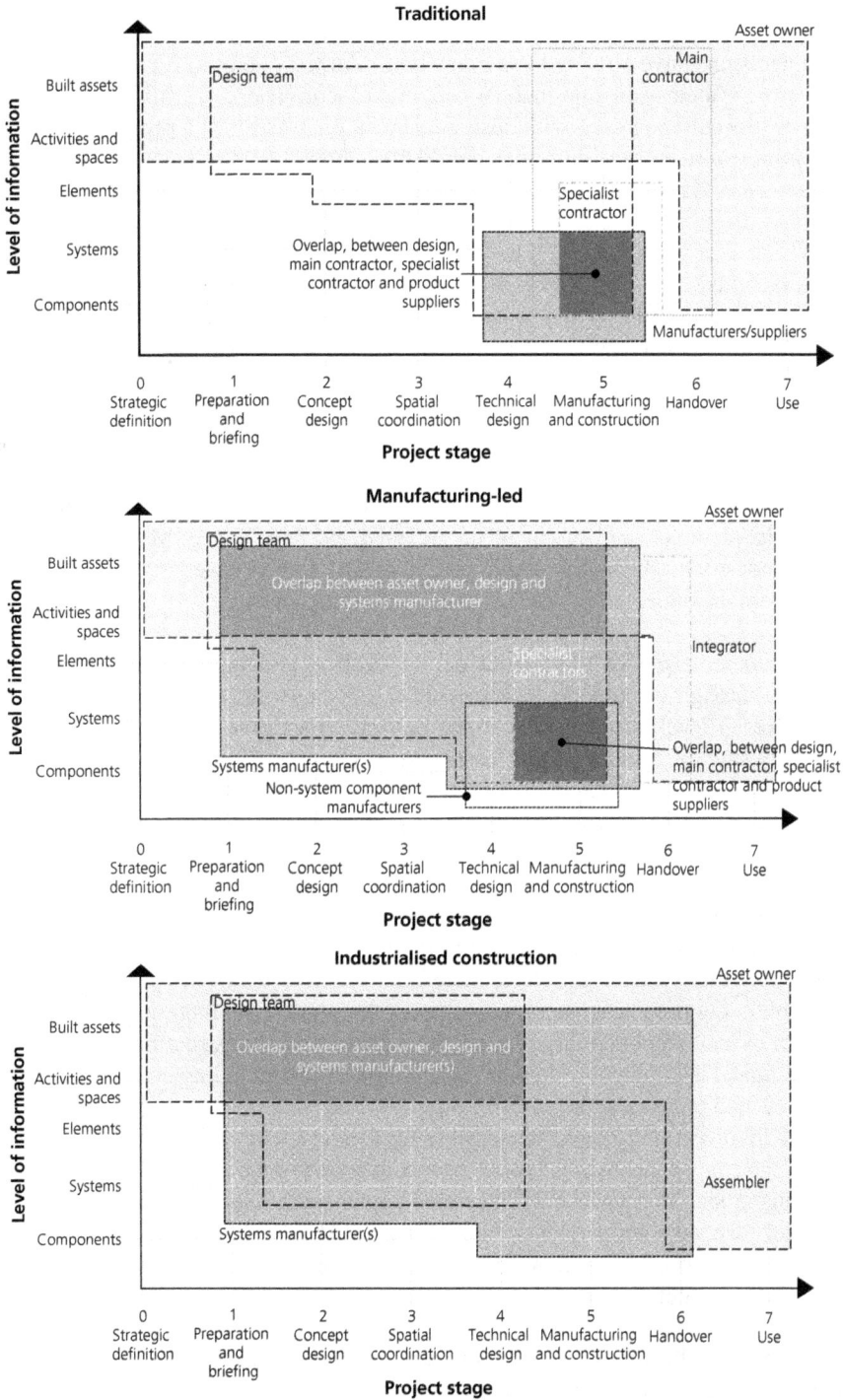

39

project-wide desired outcomes. Crucially, both main contractors and specialists are not part of the development process at the early stages of a project, and so have limited opportunity to innovate or provide valuable input or alternative solutions. Likewise, product suppliers typically have no direct contact with the client or design team other than at the technical design stage, when decisions have already been made that limit the potential for alternative solutions involving a more systems-led approach. Where manufacturers are operating at the component level, it can be difficult to engage with design teams or contractors at a project or even systems level, especially where those discussions would cross boundaries of packages already defined by the project team. As a result, opportunities for potentially value-adding improvements can be missed.

Moving to a manufacturing-led approach changes the relationships between different parties in several ways. First of all, the system manufacturer is likely to be engaged early in the project process, which enables the benefits of a manufacturing-led approach to be clearly understood and potential interfaces to be managed. Front-end loading a project (i.e. spending time during the early stages to plan, develop, test and finalise solutions) can add significant value. Of more than 20 000 projects assessed, those that were effectively front-loaded had on average 20% lower costs and were delivered 10 to 15% quicker than average (ICE, 2020). In a manufacturing environment (for example, in shoe manufacturing), the product isn't made until its user requirements are fully defined, its design complete and supply chain prepared and fully costed. Manufacturing-led construction, and industrialised construction in particular, should be viewed in the same way if they are to offer maximum value.

Systems providers are also likely to operate across levels of granularity – for example, from the built asset level through to individual components. This offers the design team the opportunity to consider a greater level of detail early in the process without the worry of abortive work and can lead to early sight of detailed project costs and programme. This model can still be complex as potentially large portions of the delivery can still be left to more traditional approaches. There will of course be less interfaces as some will be aggregated by the system provider, but the overall responsibility to deliver the project will still fall to the main contractor.

The industrialised construction model is potentially the simplest in terms of relationships but is also the furthest away from the traditional delivery model. While Figure 4.2 is simplified and suggests only one systems provider, there may be more than one, but there are unlikely to be many unless the project is large and complex, in which case there is still likely to be an integrator with overall responsibility. Where the industrialised construction model varies from the traditional model is largely through the technologies used, including customisation of standardised products and processes, the shift in responsibilities to a manufacturing-led business and the early-stage involvement of those responsible for delivery. The model is still likely to require assemblers of systems on site, which are not necessarily part of the main system provider's business. Using this model, the number of interfaces between the client and the delivery team are significantly reduced, depending on how many systems providers there are. Instead, the detailed supply chain forms part of the system provider's remit and is managed by them according to whichever delivery model they use.

The design team's role becomes more that of a configurator of systems than a detailed designer for elements covered by the selected building systems, enabling them to focus on bringing their core design skills to the project and being less concerned with the detail of aspects falling within the system provider's scope. Architects, for example, can assist in the development of configurable solutions for different sectors, enabling manufacturers to address multiple markets instead of designing one-off projects, or focus on the spatial arrangements and qualities of an asset.

The ever-increasing digitalisation of the construction industry is a key enabler of industrialised construction. Building information modelling (BIM) alone has made it easier for suppliers to cross traditionally siloed sectoral boundaries, partly because it enables solutions to be assessed based on performance rather than cultural norms alone. In addition, assets can be designed, 'built' and tested virtually many times with no risk attached, and solutions have become infinitely more visible through product libraries and platforms, such as KOPE market by KOPE.

Visibility is one of the main reasons why it may be beneficial for manufacturers to supply their products to systems providers rather than direct to projects. In some instances, it means that they can enter markets that they would otherwise not have the opportunity to enter, and in others it may mean delivering and capturing more value. McKinsey (2019) suggests that the modular construction market could reach up to 10% of infrastructure and industry spend by 2030 while delivering 10% cost savings. For a component supplier, that may provide a consistent pipeline for standard products that would otherwise have been served on a project-by-project basis with nonstandard product variations.

The industrialised construction model also provides real opportunities for small- and medium-sized (SME) businesses who may not otherwise have access to the necessary digital or physical capabilities to thrive or to operate on multiple projects but who can be supported by larger systems provider partners for mutual benefit. Open product platforms or standard component specifications may also provide opportunities for small businesses, but in a less direct way than delivering repeat business directly into a systems provider.

Technologies that enable mass customisation (the benefits of mass production combined with the benefits of customisation for specific user requirements), such as product platforms, can deliver real opportunities for system providers as the solutions can be agile to suit many sectors and projects while limiting production inefficiencies. At the same time, clients get the benefit of high-performing, mature manufactured products that are tailored to their specific requirements. It is important that, when using any form of manufacturing-led construction, the solution is not led by the capabilities of the system but that the system can be configured to deliver the required outcomes, both initially and throughout the asset's life cycle.

It is important to dispel the myth that manufacturing-led construction provides a barrier to great architecture or design, or is only applicable to a small selection of asset types. Manufacturing-led technologies and mindsets are applicable to any form of built asset type, and the clearer the definition of desired outcomes, the easier it is to ensure that the right technology is selected, instead of using the wrong solution that may negatively impact on the design of the finished asset. A good example of mass customisation can be found in the work of Automated Architecture (AUAR), described in Box 4.1.

Box 4.1 Automated Architecture (AUAR) – production and assembly flexibility

Figure 4.3 Automated architecture (Source: AUAR)

AUAR have a vision to enable the best possible way to inhabit the planet by delivering high-quality assets to the masses using an asset-light approach. The physical building components are all based around a dimensionally coordinated three-dimensional (3D) timber block. Using their purpose-built generative design-to-manufacturing and assembly suite, almost limitless design options can be created, safe in the knowledge that the designs can be produced and installed as one-dimensional (1D) components, two-dimensional (2D) panels or 3D modules, depending on project-specific access requirements. On a low-rise residential project, AUAR claim to reduce the delivery time from 36 to 6 weeks. All connections between blocks are dry, meaning that they can be disconnected and replaced or reused.

Instead of a centralised production facility, AUAR are developing an ecosystem of micro factories (a target of 40 micro-factory partners by 2027). These micro factories are themselves modular and use off-the-shelf robots to produce the timber blocks, which can then be assembled into partially-finished panels or volumes in the micro-factory and installed on site using local labour. A micro-factory requires approximately 10% of the capex of a typical off-site factory and only requires 4% of the opex to achieve the same production capacity.

Assuming that manufacturing-led solutions are a part of the solution to overcoming the many challenges facing the construction industry, there will still be a significant number of interventions that will not use such approaches. In those cases, and cases where manufacturing solutions are used, automation can also offer significant advantages over traditional ways of working.

4.3.   Automation

Automation itself is nothing new, but with the rapid development of technology in the last few decades, and its continuation in the years to come, more and more tasks may become automated, whether physical or digital tasks. PricewaterhouseCoopers (PWC, 2018) have identified three potential waves of automation leading up to the mid-2030s, based on the widespread availability of technologies, not necessarily market application as that can be driven by many factors. The three waves are

- *algorithm wave* (early 2020s) – focused on simple computational tasks and analysis of structured data
- *augmentation wave* (late 2020s) – automation of repeatable tasks, such as analysis of unstructured data, robots in warehouses and dynamic technological support
- *autonomy wave* (mid 2030s) – automation of physical labour and manual dexterity, problem solving in dynamic situations.

PWC (2018) estimate that such automation technologies could contribute as much as 14% to global GDP by 2030 across all industries, which is more than the core construction industry currently contributes. While this will have a significant positive impact on productivity, it will in the short term have a real impact on the labour force. In the long term it is widely expected that the number of jobs overall will increase as a result of automation technologies, but in the shorter term it is likely that many jobs will be significantly altered as a result. That isn't to say that entire jobs will be replaced by computers or robots, but that roles will change and adapt. Figure 4.4 illustrates the potential share of jobs that PWC suggest could be automated in construction and other industries in each of the waves.

The level of potential automation depends on a number of factors, but one method of assessing the implications is to look at the types of tasks that are undertaken in different roles. Figure 4.5 illustrates the share of tasks by type for roles identified as being within the construction industry. These splits give a reasonable indication of the potential level of automation within the industry.

Figure 4.4 Share of jobs that can be automated (Source: PWC, 2023)

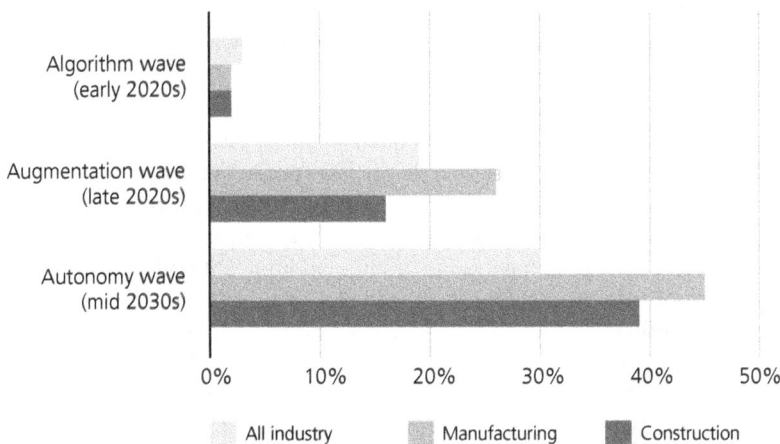

Figure 4.5 Composition of tasks by type for construction jobs (Source: PWC, 2023)

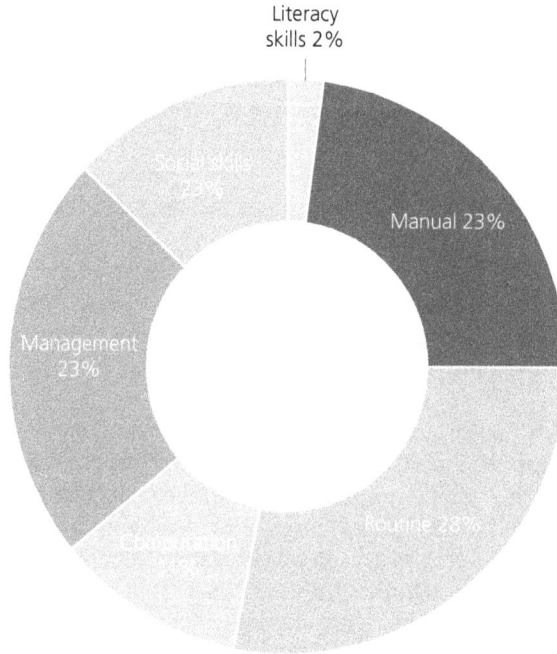

As was identified in Chapter 3, it is important to remember that the construction industry as defined for statistical purposes typically only relates to site activities, and so excludes professional services, construction product manufacturing, asset maintenance and a significant proportion of manufacturing-led construction businesses. For this reason, coupled with the range of solutions covered in this book, Chapters 22 and 23 in particular will look at the impact on the wider industry in more detail. However, it is fair to say that automation is likely to have a significant impact on the way that built assets are designed, delivered, operated and maintained.

4.4.   Summary

Automation and manufacturing-led construction are in some ways potentially doing the same thing; they are both solutions to improve on how activities can be carried out more effectively, more efficiently, safer and at greater volume with less resource. Both also include digital as well as physical aspects and both can improve the quality and consistency of output. Production of manufacturing-led construction will almost certainly incorporate automation of some sort, whether it be machines on a production line or generative design in the form of configurators and communicating designs directly to production equipment. However, not all interventions will utilise manufacturing-led construction, and the existing built environment still needs to be operated and maintained where automation technologies can add real value in terms of surveying, monitoring performance and controlling internal environments. All these areas will be covered in Chapters 15 to 20 of this book. However, a prime example of how automation and manufacturing-led construction can work together to deliver added value is with generative design, which is described in detail in Chapter 17. Referring back to Figure 4.2, imagine fully detailed, costed, compliant and resolved designs being generated in minutes instead of weeks, and the ability to select a preferred option from many that the supply chain already has in place to deliver; that is the power of advanced generative design systems. Ravnikar Soper Technology (RST), for example, claims to reduce project timescales (including design) by over 30% through the delivery of fully resolved designs, while Archi.Con.Des (UK) Limited believe that when combined with a manufacturing-led approach to delivery, the total project timescale can be reduced by 45%.

Where automation and manufacturing-led construction are used together to deliver industrial-strength supply chains capable of delivering multiple solutions with the same systems, this is industrialised construction. Industrialised construction is as close as the construction industry can get to achieving a full production system for the delivery of assets, similar to other manufacturing industries; lessons that can be learned from other industries are described in the next chapter. An industrialised built environment takes automation and manufacturing one step further, the equivalent to providing replacement or upgraded parts for a car through the use stage in its life cycle.

REFERENCES

HMG (Her Majesty's Government (2020) *The Construction Playbook*. The Stationery Office, London, UK.

HMT (Her Majesty's Treasury) (2017) *Autumn Budget 2017*. HMT, London, UK.

IPA (Infrastructure and Projects Authority) (2021) *Transforming Infrastructure Performance: Roadmap to 2030, Infrastructure and Projects Authority*. IPA, London, UK.

Inside Housing (2022) MMC cut carbon emissions by up to 45%, research finds. https://www.insidehousing.co.uk/news/mmc-cut-carbon-emissions-by-up-to-45-research-finds-75970 (accessed 15/03/2023).

ICE (Institution of Civil Engineers) (2020) *A Systems Approach to Infrastructure Delivery*. ICE, London, UK.

Mace (2021) *The New Normal*. Mace, London, UK.

Make UK (2022) *Greener, Better, Faster: Modular's Role In Solving The Housing Crisis*. Make UK, London, UK.

McKinsey (2019) *Modular Construction: From Projects to Products*. McKinsey & Company, New York, NY, USA.

McKinsey (2020) *The Next Normal in Construction*. McKinsey & Company, New York, NY, USA.

PWC (Price Waterhouse Coopers) (2018) *Will Robots Really Steal Our Jobs?* PWC, London, UK.

PWC (2023) The impact of automation on jobs. https://www.pwc.co.uk/services/economics/insights/the-impact-of-automation-on-jobs.html#data-explorer (accessed 02/04/2023).

emerald PUBLISHING   ice

Steve Thompson
ISBN 978-1-83608-599-7
https://doi.org/10.1108/978-1-83608-598-020241007

Chapter 5
Learning from other industries

5.1.   How does construction compare to other industries?

As mentioned in earlier chapters, the construction industry is recognised as one of the least digit-ised, least productive industries. Figure 5.1 presents the relative level of digitalisation and potential for automation of different industries according to McKinsey (Gandhi *et al.*, 2016, McKinsey, 2017), along with an indication of their contributions to output (ONS, 2022) and type (product, resource or service).

Figure 5.1 suggests that most of the industries with product as their primary output (including con-struction, where the product is a built asset) tend to have a high potential for automation compared to generally more knowledge-based industries with service outputs. However, it also suggests that traditional construction has the lowest potential of the product industries. Based on technologies available in 2016 when the data used in Figure 5.1 refers to, that may have been the case.

There is no widely used metric to enable comparison of levels of automation across industries, but the metric commonly used to assess the use of robotics is robot density. Across all industries globally in 2020, the average robot density was 126 robots per 10 000 workers (IFR, 2021). That is almost double the figure in 2015 (66). However, the UK is only ranked 24th in the world for robot density, with only 101 robots per 10 000 workers. In construction in 2017 the density was only 0.5 robots per 10 000 workers, so significantly below the UK average and below the European con-struction average of 1.5 (Statista, 2018). However, as has already been mentioned, robotic density is not necessarily a clear representation of how automated an industry is; physical automation can also be achieved through the use of machinery that is not classified as robotics, or through the use of nonphysical task automation.

Today and in the years to come it is clear there is significantly greater opportunity for automation in construction than previously thought, which is why the future construction bubble shown in Figure 5.1 is positioned to highlight a more realistic position understood today. There are three main reasons why previously there may have been seen to be limited potential for automation in construction relative to other industries. These are

- type and variety of product delivered by the construction industry
- method of delivery of the product
- fragmentation of the construction supply chain.

Firstly, the type of product delivered: a built asset. The range and complexity of built assets deliv-ered by the construction industry is far broader than in any other industry; assets are traditionally usually seen as being unique, even when based on a standard product such as standard house types. With such a range of products to be delivered, looking through a traditional lens it can appear to be difficult to automate the assembly of disparate products to create an almost unique finished

Figure 5.1 Level of digitalisation and potential for automation by industry (Source: Gandhi *et al.*, 2016; McKinsey (2017)

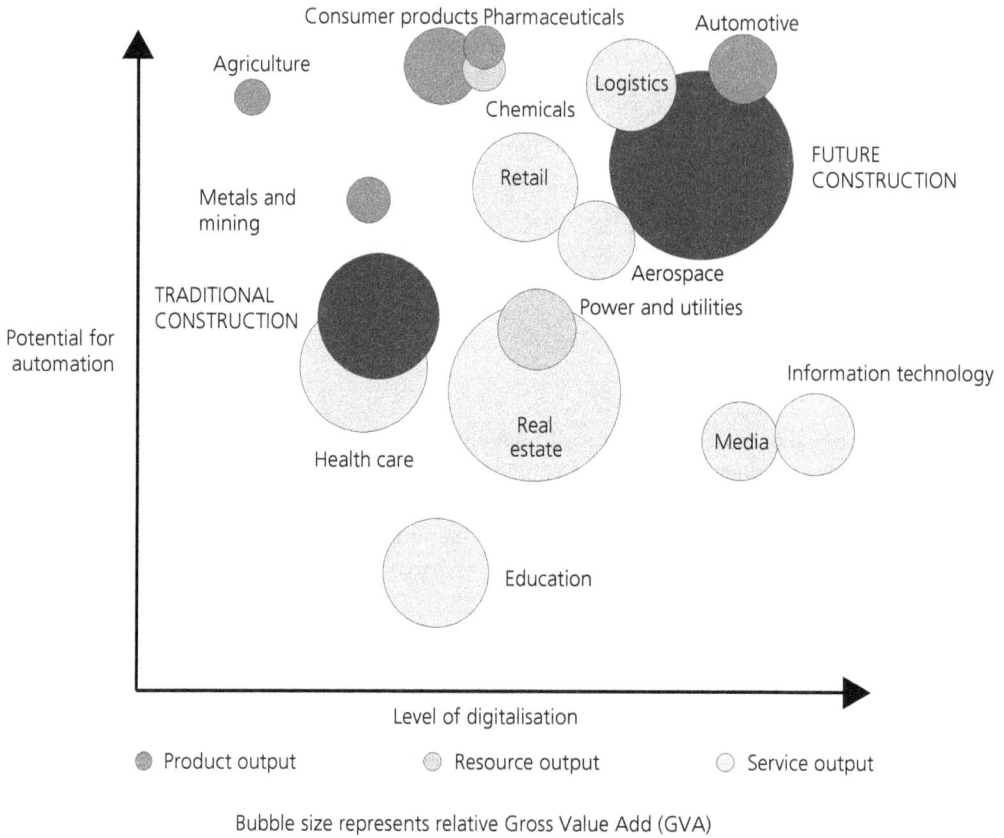

product, in unique and typically hard to control environments by teams formed specifically to deliver a particular instance of a product. The situation cannot be any harder to address, but as this book describes, when the problem is viewed with more of a manufacturing-led mindset and a focus on delivering outcomes, then the potential for mass customisation to deliver the range of end products is possible, using standardised and automated processes.

When comparing products across different industries, the supply chain operations reference model (SCOR) focuses on processes within a supply chain and identifies three main delivery methods, which are

- *made to stock* – the product is made without a specific customer in mind and without an order being placed
- *made to order* – the product is made following an order being placed by a specific customer
- *engineered to order* – the product is made to meet the specific requirements of a customer who has placed an order.

Figure 5.2 Relative repeatability of solutions by delivery method and industry (Author's own)

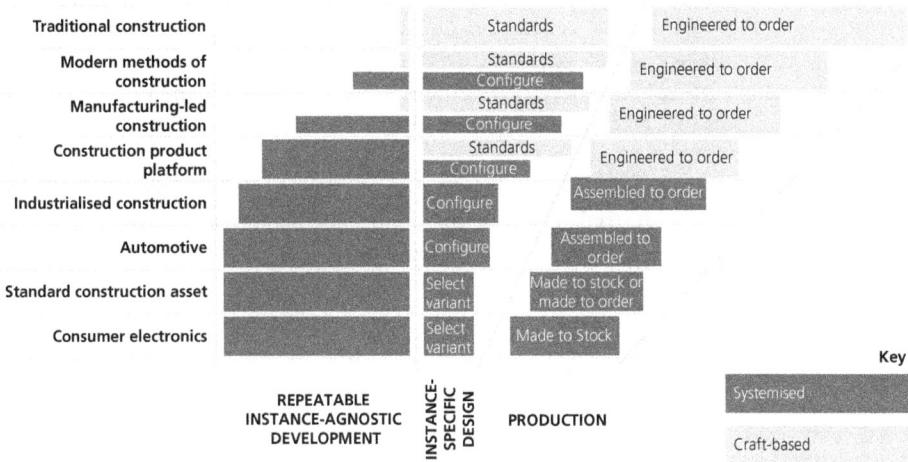

When moving from made to stock to engineered to order, not only is the relationship between the manufacturer and the customer more remote, but also the potential variations increase. Except for the automotive industry, which typically works on a made to order model, the product industries shown in Figure 5.1 generally make to stock. Traditional construction, however, is very craft-based and so at best can only be described as engineered to order. Instance-specific design may include sizing elements for specific site constraints or user requirements. As Figure 5.2 illustrates, the repeatable elements of traditional construction projects are minimal; not just because of the complexity and variation in the end product but also because of the risk associated with investing heavily in instance-agnostic development without clear visibility of a potential pipeline of sales or routes to market.

It is clear that with a move towards manufacturing-led or industrialised construction, the potential for automation increases as the level of repeatable instance-agnostic development work increases. Solutions are developed once and then applied over many instances.

Another factor that increases the potential for automation and manufacturing-led construction is the ever-increasing performance requirements for the built asset; to achieve better performance consistently, new products and processes are required. The new performance challenges also mean that automation is increasingly commercially beneficial.

5.2.   Challenges facing the construction industry

As described in Chapter 3, the construction industry typically has a strong focus on delivery of assets (the products) through projects, which are discontinuous. Manufacturing businesses, on the other hand, involve continuous processes and relationships. One issue that the project focus brings is that of the design freeze; in other industries, it is almost unheard of that a product would begin to be manufactured without its design and configuration being finalised, whereas in some areas of construction it is the norm. By limiting the amount of preproject work carried out before planning and funding have been signed off, for example, everything after those points can become squeezed to deliver an asset within as short a period as possible. That can lead to designs not being frozen until after construction work has begun, which is clearly not optimal and can lead to rework or additional cost and time.

Other industries were not always as well organised, however, with ship building, commercial aeroplane manufacturing and automotive all examples of those that have transitioned from fragmented supply chains delivering one-off products through projects.

McKinsey (2020) identified nine areas of development that an industry is likely to go through, over two waves of development, to industrialise. In the first wave, they identified

- a shift to a product-based model
- control of the value chain
- investment in technology and facilities
- investment in human resources
- customer centricity.

In the second wave, they identified

- specialisation
- consolidation
- internationalisation.

Across both waves is an increased focus on sustainability. Put simply, the first wave is about enabling a new future and setting the vision. The second wave is about optimising the industry to better deliver that vision and scaling. The big difference that the construction industry faces compared to other industries is the pace of change. Typically, the first wave has taken 20–25 years, and the second wave 10–15 years, with some overlap of the first wave. In construction, the majority of both waves is expected to occur within 5 years.

With a move towards an increasing focus on products and processes, it becomes ever more important to look at the end-to-end process of the delivery of an asset, which means developing across previous functional boundaries to create optimised solutions. For example, work by a particular team can be optimised to work with that of others instead of trades having to deal with issues arising from the earlier installation of solutions by previous trades. This focus on processes also addresses the third issue impacting the potential for automation, which is fragmentation within the supply chain. Focusing on end-to-end processes instead of function-specific trades, the potential for optimisation can be significant, and beneficial to all of those involved in delivery. It does not mean that the supply chain needs to be fully integrated but that its activities should be clearly defined and optimised.

With the pace of change that is expected to occur over the next 5 to 10 years, it is important to recognise that while lessons can be learned from industries that were previously like construction, industries such as information technology are also worth studying as they are agile and capable of dealing with constant change.

Historically, many industrialised industries have been dominated by a small number of large multinational suppliers with a dominant business model. The challenge for such incumbents comes when customer requirements change, or disrupters enter with a new model that completely changes the way the customer perceives value. Businesses may be used to dealing with one or two similar competitors and have invested heavily in production facilities to deliver products in a certain way. It can therefore be incredibly difficult to change course quickly to deal with a new, very different threat that potentially challenges the whole business model. One example is the move to online shopping with companies such as Amazon. Whereas previously anyone wishing to sell products at any great scale would typically have a physical presence, such as shops, and storage for stock and access to these would act as a barrier to new entrants. Today however, whole businesses are operating with stock stored in Amazon's warehouses and sales managed automatically through their website. Again, evidence from more agile industries such as information technology suggests that it can be beneficial to have more than one business model at play, to minimise cannibalising existing revenue streams and avoid becoming over-invested in a single model that may no longer be valid as customer requirements or preferences change. The wider construction industry needs to consider developing a continuous development mindset to cope with ever-changing requirements, instead of developing new piecemeal business models that themselves may quickly become redundant as new technologies become embedded or customer requirements continually change. The following chapters will look at how both physical and digital technologies can be used to provide flexibility in how value can be delivered and taken.

With the rapid development of technologies over recent years and expected in the years to come, it is important that the use of technology does not become self-fulfilling. For example, over the last decade there has been a significant shift by much of the industry towards using building information modelling (BIM) on construction projects. While this shift has significantly changed parts of the industry, in many cases it has been used for the sake of using it, not for adding real value to the delivery team or client. When adopting new technologies or processes, it is essential that their use is purpose driven, and that they are considered as part of a toolkit, not as results in themselves. As the industry becomes more technologically enabled it becomes more important than ever to ensure that it considers the human impact, both in terms of capability but also outcomes. One of the main differences with the technologies that are becoming available now is that previous industrial revolutions have largely involved enhancing our physical capabilities, whereas now technologies can potentially augment our mental abilities. Allowing that to happen unchecked can lead to significant consequences; for example, what data is being used to train AI systems that determine whether an autonomous vehicle should turn or break to avoid an obstacle? To ensure the focus remains on delivering required outcomes, it can be useful to take a systems approach to the problem.

5.3. Systems engineering

Systems engineering has been around in different guises since the beginning of the 20th century and has been used as an approach to delivering complex developments across many industries since, including automotive, pharmaceuticals, aerospace and defence and other manufacturing industries.

Figure 5.3 illustrates the systems 'V', a common diagram used to illustrate the systems engineering model. Teams begin at the top left by developing a big-picture concept of the desired outcome – for example, the requirements for an operational aeroplane, which may be to carry a certain number of passengers a specified distance within a given timescale. From these high-level requirements, it works its way down the left side of the V from top to bottom and, as it does so, a system architecture is created that describes how the outcome will be achieved. In doing so, the level of detail of requirements increases as it passes from one stage to the next. One of the main benefits of a systems engineering approach is that it is very rigorous at dealing with requirements definition and driving

Figure 5.3 The systems 'V' (Author's own)

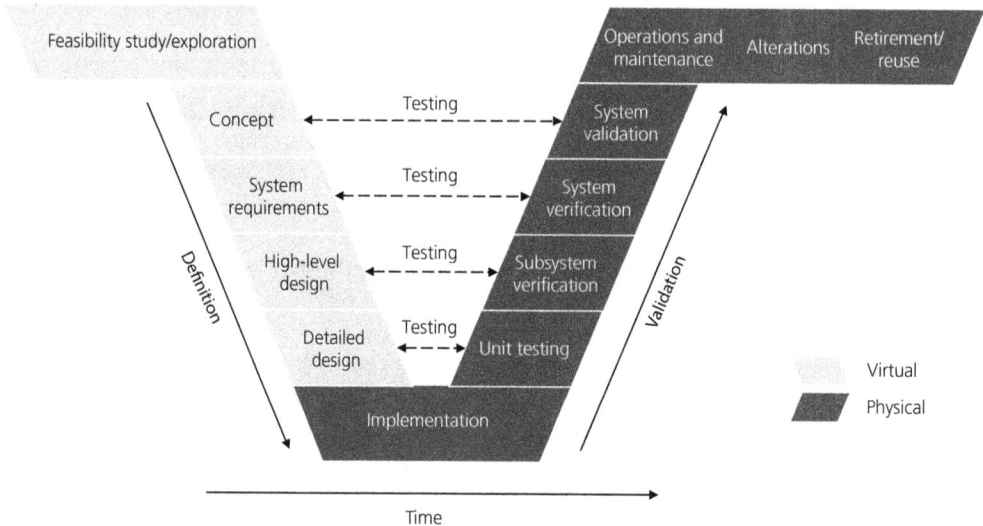

big-picture requirements down into requirements for systems, subsystems and components while ensuring that all requirements are linked to those at a higher level.

Once the requirements have been thoroughly defined through the left side of the V, implementation occurs at the bottom, which is where outputs move from purely virtual to a combination of physical and virtual. Then the process moves up the right side of the V through testing of individual components and systems, through to verification and validation at different levels of granularity until the output is delivered, operated and maintained. It is important to note that outputs are not only physical but also include other systems that are required to deliver the necessary system of systems. For example, an aeroplane needs the software and capability to communicate with air traffic control systems to be able to operate, the delivery of a physical plane alone is not sufficient. In the same way, the successful operation of physical infrastructure relies on the use of significant data and technology that need to be considered as part of the system of systems.

As the linear process moves from one stage of the V to the next, documents are exchanged to support the next step. Once outputs are delivered, they are tested against the requirements set earlier on the left of the V, at the appropriate level of detail before moving to the next stage. This links back to Figure 4.2, where requirements need to be met before moving to the next level of granularity. One of the beauties of the systems engineering approach is the ability to move between different levels of granularity to understand how requirements and solutions are linked, and to ensure testing occurs at the appropriate level through the life cycle of the project.

A systems approach does require significant effort up front but, particularly when dealing with very complex interventions or developing new systems for multiple projects, the added value can significantly outweigh the additional initial effort. In their 2022 review *Systems Approach to Infrastructure Delivery* (SAID) (ICE, 2022), the Institution of Civil Engineers identified the potential for a systems approach.

While the traditional systems engineering method is extremely thorough and effective, the fact that it is a linear, document-led approach can be restrictive and can potentially lead to errors being hidden within documents that are difficult to access. A new approach called model-based systems engineering (MBSE) has developed that can provide the benefits of a traditional systems approach with the added value of instant access to the right information.

5.4. The move to model-based systems engineering (MBSE)

As the construction industry has seen a move towards BIM and federated models and, as with many other industries the development of digital twins, it should come as no surprise that delivery of systems engineering has also moved into the realms of real-time models. In simple terms, MBSE can be seen as an agile approach to systems engineering which relies on shared access to a common, up-to-date model of the latest data on a system's development. Figure 5.4 illustrates that MBSE no longer relies on a linear process such as the systems V, but instead is centralised and simultaneously links to a number of activities.

In MBSE the full set of requirements of a system are linked in a database where each is an individual object which forms part of an overall configuration. This means requirements are linked to each other and to parts of the developing system. By making these requirements model-based, a digital thread can be created that runs through the system life cycle and reduces blind spots that may otherwise occur. At the same time, the move to a model-based approach means that different

Figure 5.4 Model-based systems engineering (Author's own)

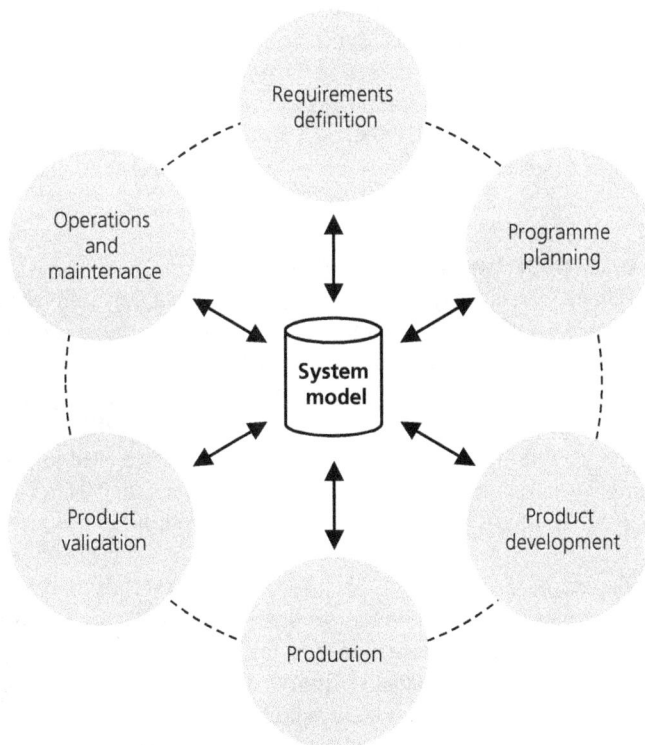

domains can work concurrently, thus removing silos and enabling teams to predict performance and virtually test and optimise the system under development.

Siemens (Kaelble, 2022) report that using MBSE can reduce the development cycle by as much as 30%, while also providing the ability to quickly change focus from the big picture to the finest of details, clear in the knowledge of what the relationships are between the two. With the digital thread that is created, tasks can more readily be automated and functions connected, meaning that sharing and learning from data in the model can be improved. To be able to benefit from this automation, however, it is critical that the systems development is process-driven, not domain-driven. In construction this means developing solutions based on the end-to-end processes required to deliver a system, not based on siloed development by profession or trade.

Working with federated and validated models in a virtual environment means that they can effectively become virtual test beds, reducing the need for physical prototypes and testing to some degree. The benefits to a product manufacturer, for example, can include the ability to see their product or system in use and to understand how it performs as part of a wider system. It is not uncommon for product manufacturers to know everything there is to know about their products and how they perform under laboratory conditions but to have little to no knowledge of how they perform in practice in a built asset. Using a modelling approach (not just geometrical modelling) can enable direct feedback to the supplier that can lead to future product development.

A further benefit of the MBSE model against the linear system V is that the digital thread includes all data from the entire programme. For example, if a component or system has been verified in a certain configuration to show compliance, evidence is retained, as are the processes used to achieve compliance. Using the Five Ps model described in Chapter 3, it is easy to compare different combinations of resources virtually through the model to find the optimum solution at a component, system and asset level, and the implications of each scenario. And on completion of a built asset, a digital twin is available that contains a full record of all the decisions that have been made in developing the final solution.

5.5. A move to more agile delivery

Most other industries have seen significant changes in the way that they deliver value to customers over recent decades; an obvious example is the impact Uber has had on private transport, utilising their platform to leverage drivers' own vehicles to provide a flexible and responsive service. To meet the challenges ahead and to be responsive to evolving client expectations and market demands, the construction industry needs to change how assets are delivered and maintained. As the product changes to meet new requirements, so must the processes used to deliver and maintain it. In Japan, for example (where off-site construction is well established), after-sales services have developed to provide whole-house warranties to differentiate suppliers.

With the rapid development of technology that the world is currently experiencing, traditional barriers to entry for new entrants are dropping. As described in Chapter 14, some businesses are developing technology-led off-site construction businesses without their own production facilities, meaning they don't need the significant upfront investment necessary for a new factory and enabling them to benefit from being more agile to suit changing requirements. Production facilities will always be required somewhere in the supply chain; it is just a matter of where they are, what they produce and how flexible they are to adapt production to suit changing requirements.

As described in Box 5.1, DataForm Lab work with clients' manufacturers to optimise automation of their production facilities to efficiently deliver systems and their variations to meet their business objectives. It is not just about having an efficient production line to deliver one type of product, there also needs to be flexibility to cope with different scenarios.

Box 5.1 DataForm Lab

DataForm Lab are a construction tech start-up focusing on the off-site construction sector. They work with off-site system suppliers, offering clear design and manufacturing automation strategies and helping them simulate and evaluate the implications of the integration of industrial robotics and automation technologies on design and production processes. This ensures production efficiency and manufacturability of systems prior to investment.

Using systems engineering logic, DataForm Lab assess and develop solutions to link design to fabrication and to optimise production facilities to provide the flexibility and productivity required. It is not just about providing the most efficient production line to deliver a particular system but also ensuring that production can adapt as required to produce different system variants with minimal impact to the manufacturer and their clients. Factory production simulations and use of a design-to-manufacturing configurator help customers understand and test their potential workflows before investing significantly in potentially the wrong automation solutions.

DataForm Lab have also developed the Auto/Mate platform described in Box 17.4 in Chapter 17, a B2B platform for off-site construction that automatically translates design information into manufacturing drawings and code, while allowing manufacturers to optimise their factory operation. Figure 5.5 illustrates the principles of automation for off-site by DataForm Lab.

Figure 5.5 DataForm Lab automation for off-site construction (Source: DataForm Lab)

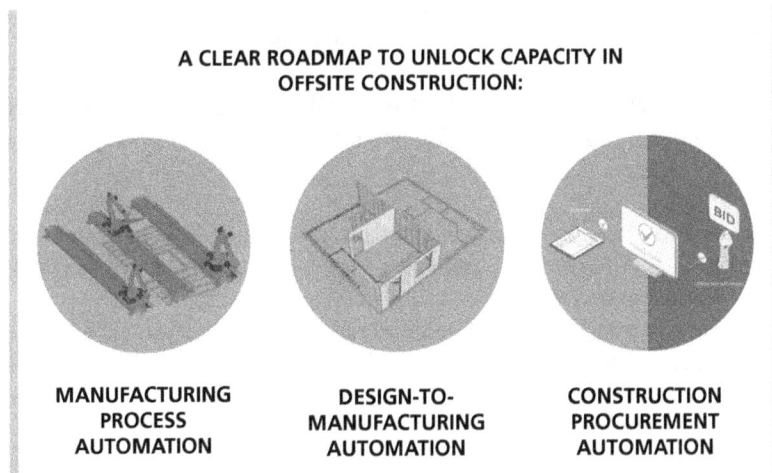

A CLEAR ROADMAP TO UNLOCK CAPACITY IN OFFSITE CONSTRUCTION:

MANUFACTURING PROCESS AUTOMATION   **DESIGN-TO-MANUFACTURING AUTOMATION**   **CONSTRUCTION PROCUREMENT AUTOMATION**

5.6. Robotics and physical automation

What is a robot? Well, in its broadest sense, robots can be described as machines that can be programmed to perform activities such as movement, manipulation or positioning. Robots used to undertake activities on construction sites will be covered in later chapters, whereas in this section the focus is on automation within production facilities, which includes industrial robots. These are the types of robots typically found in factory environments and are defined in BS ISO 8373:2021 as

> automatically controlled, reprogrammable multipurpose manipulators, programmable in three or more axes, which can be either fixed in place or fixed to a mobile platform for use in automation applications in an industrial environment (BSI, 2021).

As mentioned earlier in this chapter, robots are used widely in other manufacturing industries, significantly more than in the construction industry. Obviously one of the main reasons for this is that they typically need a controlled environment and are not usually easily transferable to individual construction sites. However, with a move to manufacturing-led construction there is greater potential for industrial robots than ever to help deliver off-site building systems. Still, not all automation in a production facility is in the form of robots; there are also nailing bridges, lathes and computer numerical control (CNC) machines, for example. The main difference between robots and CNC machines is that CNC machines are very accurate at performing one task, whereas robots can perform many tasks. It is this flexibility that makes robots appealing, as they are more able to cope with reconfiguration of production lines to deal with changes in requirements.

In a 2022 study by McKinsey (2022), respondents to a survey from several industries (not construction) identified improvements in output quality, reliability, speed and capacity as the main reasons for automating production, with cost being by far the greatest barrier, followed by lack of internal experience. The most common use cases for automation in production today were identified as picking, packaging and unloading, and for future automation were material handling, sorting and quality assurance. While robots are not typically used in construction manufacturing to date, these potential use cases are relevant to the industry, and increasingly so.

5.7. Digitisation, digitalisation and digital transformation

Digitisation, digitalisation and digital transformation are terms that are often confused and used interchangeably. Put simply

- *digitisation* refers to changing something from analogue to digital – for example, scanning a paper document or entering product data into a spreadsheet or online tool so that the data can be accessed and used more readily
- *digitalisation* uses data and digital technologies to improve processes. Examples include using product data to determine expected performance of a wall construction instead of calculating manually or tracking progress of a product delivery
- *digital transformation* is business transformation enabled by digital technologies – for example, predictive maintenance of assets instead of responsive maintenance when something has already gone wrong.

Each of them is relevant to the future of automation and manufacturing-led construction and each is reliant on the previous one.

5.7.1   Digitisation

Digitisation is the bedrock of digitalisation, digital transformation and automation. Without data, none of those would be possible, which is why it is crucial for businesses operating in construction to digitise their product and service data as soon as they can, to benefit them and their customers. This will be covered in some detail in Chapter 10. Automation can be used to support digitisation by measuring, recording and capturing data from analogue sources – for example, through laser scans or drone surveys, or simply through image scanning and computer vision.

5.7.2   Digitalisation

Digitalisation is about harnessing data from across the product life cycle and across all relevant domains to enable better decision making. This can include all levels of granularity – that is, components, systems, built assets and individuals, businesses and communities. Digitalisation is crucial to modelling and visualising the complexity of a construction intervention and analysing outputs, and usually involves some form of calculation. Most of the technologies described here, whether automation or manufacturing-led, either rely on or directly involve digitalisation in some form.

5.7.3   Digital transformation

Digital transformation is about finding new and better ways of doing things, but also about ensuring the right things are done. It is often outcomes-driven – that is, focused on what the customer really wants or needs and creating new ways to deliver that, supported by digital technologies. It typically requires an end-to-end approach, and actually one of the most common failures of digital transformations is where new digital capabilities are bolted on to existing practices that do not fit well as part of a changed model. It can include the use of digital twins, generative design or other forms of artificial intelligence (AI). Again, many of the technologies described in this book can form part of a digital transformation, but the transformation needs to be driven by the requirement to deliver certain outcomes, not just by a desire to use new technologies without the necessary business strategy.

5.8.   Digital twins – merging the physical and virtual worlds

Digital twins have been a hot topic in parts of the construction industry in recent years, as they have in other industries. However, they have been around in some forms for several decades. The term digital twin is thought to have been coined by Michael Grieves in 2002, but NASA were using basic digital twins in the 1960s.

Figure 5.6 describes the difference between digital models, digital shadows and digital twins. It is important to note that digital twins are not necessarily twins of physical objects but can also be twins of processes, which means the benefits of such approaches can be realised far wider than twins of physical assets, for example in procurement.

The Department for Business, Energy and Industrial Strategy (BEIS) identified three core areas where cyber-physical systems, for example, digital twins, can deliver benefits (BEIS, 2022). These are

- *design, operation and optimisation* – real- or right-time information can inform design, test and refine products and services, including operation of smart robots and unmanned vehicles
- *strategic planning and scenario modelling* – accurate and up-to-date models of a product or process can support accurate planning and future scenario modelling
- *resilient, flexible and responsive systems* – responding to rapidly changing requirements.

Figure 5.6 Digital models and digital twins (BEIS, 2022)

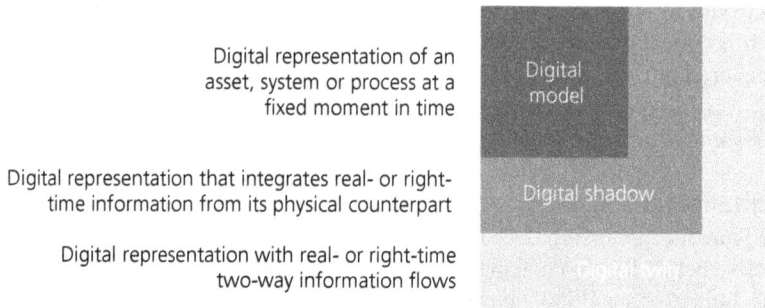

Digital representation of an asset, system or process at a fixed moment in time

Digital representation that integrates real- or right-time information from its physical counterpart

Digital representation with real- or right-time two-way information flows

One example use case of digital twins is the production digital twin, where real-time models of production facilities are created. These can integrate computer-aided design (CAD), computer-aided manufacturing (CAM) and Internet of Things technologies (IoT) with the physical assets of a production facility. Real-time data from machines can then feed into the model to improve decision making and ultimately improve production efficiencies and performance.

5.9. Virtual automation – artificial intelligence

Artificial intelligence (AI), like digital twins, has been around for some time, but capabilities have developed almost exponentially over recent years. ChatGPT is one of many tools that has gained widespread interest from across industries due to its potential to change the way AI can be accessed and used. The use of AI in construction will be described in more detail in later chapters, such as in generative design or predictive maintenance. However, first this section provides a high-level overview of AI and its different forms.

Artificial intelligence is a broad term that covers several technologies and can be broadly described as the ability of a machine to imitate human intelligence. Some of the key technologies associated with AI are

- *machine learning* (ML) – algorithms are used to process large volumes of data and learn from it. Once taught, ML models then make predictions or identify patterns without being programmed to do so and, as such, are more advanced than simple rules-based models. There are different types of ML with varying reliance on human training. These are supervised ML, unsupervised ML, reinforcement ML and deep learning, where algorithms process large volumes of unstructured data inspired by how neural networks operate
- *natural language processing* (NLP) – machines recognise text or voice, extract data and use that to produce an output
- *computer vision* (CV) – enables computers to analyse images and provide appropriate output – for example, facial recognition
- *robotics process automation* (RPA) – pre-programmed software automates labour-intensive, repetitive tasks.

5.10. Summary

To summarise, it is fair to say that a lot can be learned from other industries when looking at addressing the challenges that the construction industry faces. However, the construction industry does not have one particular industry that it is similar to; it has many that it can learn some things from. The key is to be very clear on what the desired outcomes are before developing a solution to a problem, but also to make sure that future flexibility is considered from the outset. The world is rapidly changing, as are markets and user requirements, so businesses need to be capable of changing their business models if they are to avoid becoming stuck with ever-decreasing market opportunities.

REFERENCES

BEIS (Department for Business, Energy and Industrial Strategy) (2022) *Enabling a National Cyber-Physical Infrastructure to Catalyse Innovation*. BEIS, London, UK.

BSI (2021) BS ISO 8373:2021: Robotics. Vocabulary. BSI, London, UK.

Gandhi P, Khanna S and Ramaswamy S (2016) Which Industries Are the Most Digital and Why? https://hbr.org/2016/04/a-chart-that-shows-which-industries-are-the-most-digital-and-why (accessed 03/04/2023).

ICE (Institution of Civil Engineers) (2022) *A Systems Approach to Infrastructure Delivery Main Report*. ICE, London, UK.

IFR (International Federation of Robotics) (2022) *Industrial Robotics 2022 Executive Summary*. VDMA Services GmbH, Frankfurt, Germany.

Kaelble S (2022) *MBSE for Dummies*. Wiley, Hoboken, NJ, USA.

McKinsey (2017) *A Future That Works: Automation, Employment and Productivity*. McKinsey & Company, New York, NY, USA.

McKinsey (2020) *The Next Normal in Construction*. McKinsey & Company, New York, NY, USA.

McKinsey (2022) *Unlocking the Industrial Potential of Robotics and Automation*. McKinsey & Company, New York, NY, USA.

ONS (Office for National Statistics) (2022) *Regional Gross Value Added (Balanced) by Industry: all International Territorial Level (ITL) Regions*. ONS, London, UK.

Statista (2018) Number of Robots per 10 000 Workers in the Construction Industry in Selected European Countries in 2017. https://www.statista.com/statistics/1012227/robot-density-construction-industry-europe/ (accessed 04/04/2023).

emerald PUBLISHING   ice Publishing

Steve Thompson
ISBN 978-1-83608-599-7
https://doi.org/10.1108/978-1-83608-598-020241008

Chapter 6

The modelling framework

6.1.   The modelling framework

The purpose of the modelling framework is to provide a consistent approach to describing the potential impact of different automation and manufacturing-led approaches across the life cycle and range of built assets, and to compare solutions against existing benchmarks.

The framework is made up of three orientations

- *vertical orientation*, which describes the life cycle stages of projects, assets, products and the supply chain
- *horizontal orientation*, which describes the services required to deliver and maintain the built environment
- *perpendicular orientation*, which describes the range of built assets that are to be delivered and maintained.

6.2.   Vertical orientation

The vertical orientation of the framework is made up of four recognised models, each describing stages in a project's, product's or asset's life cycle in a different way. The reason for having more than one is simply because they each describe a different, overlapping life cycle and the benefits of using new technologies or processes to manage or deliver the built environment cannot be measured during a project life cycle alone. Sometimes the value is realised across multiple projects or products, within the supply chain or through the operational phase of a built asset.

The first model included in the vertical orientation is the RIBA Plan of Work 2020, published by the Royal Institute of British Architects (RIBA 2020). The model covers the briefing of construction *projects* through to the operation of the finished built asset; however, the focus is on the design and construction phases of a project. The project life cycle according to the RIBA Plan of Work 2020 includes

- 0 Strategic definition
- 1 Preparation and briefing
- 2 Concept design
- 3 Spatial coordination
- 4 Technical design
- 5 Manufacturing and construction
- 6 Handover
- 7 Use.

The second model in the vertical orientation provides a *value* perspective of a built asset and is defined in the Construction Innovation Hub Value Toolkit (Construction Innovation Hub, 2023) and BSI Flex 390 (BSI, 2023). This model has been used to describe the potential benefits of new technologies and processes through the asset life cycle to the asset owners, occupiers and wider society. The key asset life cycle stages are

- need
- optioneering
- design
- delivery
- optimisation.

The third model in the vertical orientation covers the life cycle of *products and systems* within the built environment. The model used is from BS EN 15804 (BSI, 2019), which is the standard used as the basis for product environmental product declarations (EPDs). Historically, many models that have been used to illustrate the value of different approaches to construction have not covered the impact on the supply chain that can be shared across projects. However, as the industry potentially moves to a more manufacturing-led model, this framework provides a real opportunity to describe the benefits of such a shift, especially benefits that impact products across many projects.

The stages of the product model include

- Product stage
    - o A1 Raw materials
    - o A2 Transport
    - o A3 Manufacturing
- Assembly stage
    - o A4 Transport
    - o A5 Assembly
- Use stage
    - o B1 Use
    - o B2 Maintenance
    - o B3 Repair
    - o B4 Replacement
    - o B5 Refurbishment
    - o B6 Operational energy use
    - o B7 Operational water use
- End-of-life stage
    - o C1 Deconstruction/demolition
    - o C2 Transport
    - o C3 Waste processing
    - o C4 Disposal
- Beyond the system boundary
    - o Re-distribute/re-use
    - o Re-manufacture
    - o Recycle material.

The final model used to describe the vertical orientation focuses on the *supply chain*. As with the product model, the supply chain model looks at the different stages involved in delivering a product or service to a project or asset. This enables us to describe the value of different products and processes to the supply chain, which ultimately has an impact on the value to many assets, not just a single project. The model used is based on the supply chain operations reference (SCOR) (ASCM 2022a). The SCOR framework enables users to describe and model supply chains in detail, but for the purposes of this book, only the top level is used. The stages are

- plan
- source
- make
- deliver
- return.

It should be noted that SCOR is used to describe traditional linear supply chains and that there is a new model called the digital capabilities model (DCM) (ASCM, 2022b). The DCM describes an agile interconnected network which relies on several technologies that are not currently widespread in construction and, as a result, is not suitable to describe the benefits of new technologies and processes directly. However, the model is useful in helping to visualise what a future construction supply chain may look like, and so is relevant to the future scenarios described in Chapter 20.

Figure 6.1 illustrates the vertical orientation of the framework and the alignment with each element.

Figure 6.1 Vertical orientation of the modelling framework (Author's own)

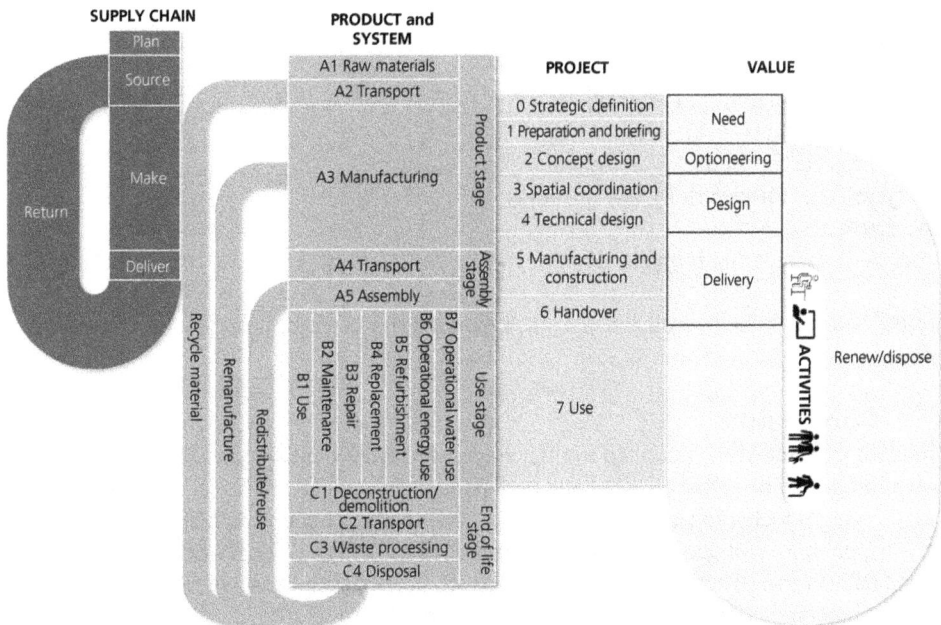

6.3. Horizontal orientation

The horizontal orientation of the framework is used to describe the value to *services* provided in delivering and maintaining the built environment. The model used is based on the standard industrial classification of economic activities (SIC) (ONS, 2022), which describes the primary activity of a business and is used widely when providing statistics on business performance across the economy. Fifty-nine industries within SIC have been identified as relevant to the built environment for the purposes of this book. For the purposes of the framework they can be summarised as

- professional services
- manufacturing and sale of construction products
- manufacturing and sale of assemblies and systems
- construction of buildings
- construction of infrastructure
- specialist construction activities
- repair and maintenance
- combined facilities support activities
- construction equipment sale and hire.

6.4. Perpendicular orientation

The perpendicular orientation of the framework defines the type of *asset* that is being delivered or maintained. It is described as perpendicular because, while construction projects deliver built assets, the value of a technology or process needs to be considered across multiple projects and asset types. For modelling purposes, the Entities table of Uniclass 2015 (NBS, 2022) has been used to define different asset types; however, for the framework, these have been grouped into categories used to present statistics by the Office for National Statistics (ONS, 2022). The categories used are

- commercial buildings
- domestic buildings
- roads and motorways
- railways and underground railways
- bridges and tunnels
- utility projects for fluids
- utility projects for electricity and telecoms
- water projects
- other civil engineering projects.

Figure 6.2 illustrates the overall framework, including the vertical, horizontal and perpendicular orientations.

6.5. Conclusion

The modelling framework described in this chapter has been developed to enable the benefits of new technologies to be better assessed and understood across different life cycle stages, assets and processes. Not all technologies or methodologies will be beneficial across all life cycle stages or asset types, but may still add value to different processes, and this framework enables them to be described and compared consistently. The framework is used as the basis for describing the impact of different automation technologies in Chapters 15 to 20.

Figure 6.2 Complete modelling framework (Author's own)

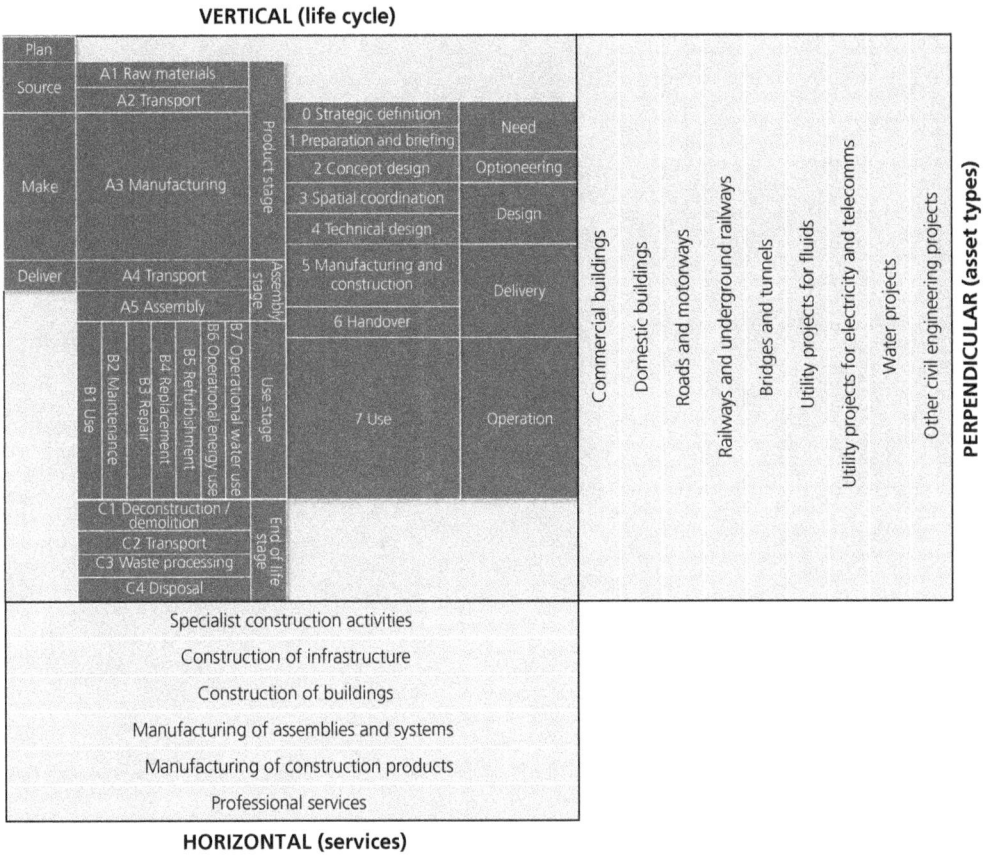

VERTICAL (life cycle)

Plan					
Source	A1 Raw materials				
	A2 Transport				
		0 Strategic definition	Need		
		1 Preparation and briefing			
		2 Concept design	Optioneering		
Make	A3 Manufacturing	3 Spatial coordination	Design		
		4 Technical design			
Deliver	A4 Transport	5 Manufacturing and construction	Delivery		
	A5 Assembly	6 Handover			
		7 Use	Operation		
	C1 Deconstruction / demolition				
	C2 Transport				
	C3 Waste processing				
	C4 Disposal				

Product stage · Assembly stage · Use stage · End of life stage

B1 Use · B2 Maintenance · B3 Repair · B4 Replacement · B5 Refurbishment · B6 Operational energy use · B7 Operational water use

Commercial buildings · Domestic buildings · Roads and motorways · Railways and underground railways · Bridges and tunnels · Utility projects for fluids · Utility projects for electricity and telecomms · Water projects · Other civil engineering projects

PERPENDICULAR (asset types)

Specialist construction activities

Construction of infrastructure

Construction of buildings

Manufacturing of assemblies and systems

Manufacturing of construction products

Professional services

HORIZONTAL (services)

REFERENCES

ASCM (Association for Supply Chain Management) (2022a) https://scor.ascm.org/practices/introduction (accessed 23/06/2022).

ASCM (2022b) https://www.ascm.org/corporate-transformation/dcm/ (accessed 23/06/2022).

BSI (2019) BS EN 15804:2012+A2:2019. Sustainability of construction works. Environmental product declarations. Core rules for the product category of construction products. BSI, London, UK.

BSI (2023) *BSI Flex 390: Built Environment – Value-based decision-making – Specification*. BSI, London, UK.

Construction Innovation Hub (2023) The Value Toolkit. https://constructioninnovationhub.org.uk/our-projects-and-impact/value-toolkit/ (accessed 16/12/2023).

NBS (National Building Specification) (2022) https://www.thenbs.com/uniclass-signpost (accessed 23/06/2022).

ONS (2022) https://www.ons.gov.uk/methodology/classificationsandstandards/ukstandardindustrialclassificationofeconomicactivities/uksic2007 (accessed 23/06/2022).

RIBA (2020) *RIBA Plan of Work 2020*. RIBA, London, UK.

Enable

emerald PUBLISHING   ice Publishing

Steve Thompson
ISBN 978-1-83608-599-7
https://doi.org/10.1108/978-1-83608-598-020241009

Chapter 7
Enabling change and manufacturing-led construction

7.1.   Introduction

The purpose of this chapter is to describe some of the conditions that can be put in place to prepare for the technological and process shift that is likely to occur in the built environment sectors over the next two decades. The focus is not on governmental or industry-wide preparations but on areas that individual organisations can put in place or consider to enable them to thrive in a rapidly changing environment, and highlights changes in mindset as much as processes or capabilities. The chapter is organised around the five steps identified in Figure 1.2 in Chapter 1, the first of which is to enable and prepare for change.

The rapid technological shift that is expected to occur along with the use of new methods of construction is likely to create the seismic change the industry needs to deliver better performing assets quicker, and with less resource (both human and otherwise). However, it is important to not just jump straight to implementing shiny new tech without first carefully considering what the purpose is, and which technologies are likely to work together well to deliver improvements. To build smarter and better (instead of just building more) requires more than just a technological shift. In fact, it is clear from other industries that have gone through significant technological change that technology application on its own is unlikely to lead to success. The industry needs to move to a people-led, technology-enabled industry instead of one led by technology alone. In other words, it is important to understand how technology can support the delivery and operation of a better built environment and how it can help businesses thrive in an ever-changing environment. That requires an understanding of the skills, relationships and cultural changes required to best work alongside technologies to deliver and operate built assets and systems.

For individual businesses it will be increasingly important to be agile, to be able to change how they operate and even potentially change markets. It is likely that, going forwards, businesses with multiple business model options are more able to succeed and be sustainable as they will not be as constrained by outside influences and will be capable of shifting between approaches as required, so a portfolio of business models may well be preferable to just relying on one. This is covered in Section 7.2 in more detail.

It will become more necessary to consider the life cycle of both products or systems and of the assets that they become a part of. It will be important to provide supporting information on how they are likely to perform, what maintenance they will require and how they can be reused or recycled at the end of their life or that of the asset, whichever is shorter. In considering such issues, new business models may well arise, such as take-back of products, product leasing instead of selling or even servitising product offerings by providing guarantees on performance and retaining the right to replace products or systems to enable that to happen. These potential new models also open up the opportunity for new relationships, including with asset owners. While likely to require different communication approaches, such relationships may lead to further longer-term and multiasset opportunities that would otherwise not be possible.

When considering both new and existing offers in a changing environment, it will be important to benchmark performance and costs of existing solutions or targets to evidence the potential of new solutions and also to ensure that they enter the market at a sustainable level with a clear understanding of the difference between what they offer compared to incumbents. Part of the comparison of solutions may also include delivery model assessments (DMA) which provide an analytical, evidence-based approach to assessing how the delivery of projects should be structured. As with considering the life cycle of solutions, if DMA is used to consider project models that are likely to be used to deliver assets or systems, this can bring new business models to light that are complementary to the technical solutions being offered. Benchmarking, procurement and supply chain considerations are discussed in Chapter 18.

Finally, especially in the infrastructure sector, it is crucial that projects be considered as interventions in systems of systems. It is increasingly likely that physical assets will support the wider application of technologies that are critical to operating transportation systems and which have shorter lifespans than the physical assets. The ability to upgrade or replace these technologies then becomes a key requirement when looking to minimise disruption over an extended period of time.

So, what can businesses do to enable them to develop and succeed in the rapidly changing technological and cultural shifts that are forecast over the next two decades? The next section looks at change management and enabling new business models supported by new technologies and construction methods that may come out of reviewing these wider requirements. Following that are sections covering the remaining stages identified in Figure 1.2.

7.2.   Business models and change management

When considering future business models and technical and commercial solutions in the context of technological change, it is crucial to look beyond the boundaries of an existing business and look at the wider supply chain, whether that be upstream and downstream operations or parallel businesses and activities. The reasons are threefold.

- The processes and services around a particular business offering, whether directly related or not, need to be considered to look for potential opportunities and barriers to change. It would be naïve to think that everything outside of the business in question will remain the same and not be open to changes in processes, technologies or even need. Whether the same offering will be required in the future or not needs to be considered as part of this.
- Analysis is required to understand whether existing or proposed partners of a new product or service have a complementary vision of the future and whether they are likely to support or limit the potential of an offer. Support may be in the form of complementary services, and limits may include competing to deliver the same proposed services.
- The risk of developing new solutions or applying new technologies needs to be considered in the wider context. For example, will a new offer impact or compete with current customer offers and thus provide a conflict with existing business and relationships that will ideally be maintained?

Traditionally, the construction supply chain adopts a functional approach (for example, packaging works of a similar type or material, such as steelwork, concrete works). This can lead to decisions being made based on individual functions, creating a silo effect. When looking at potential offers for the future, it is important to take a broader view and consider wider processes involved to provide a clearer picture of potential interfaces. Such an approach offers the potential for more productive, collaborative and less transactional relationships in the long run as it can lead to a shared understanding and appreciation of what is to be delivered by who, with risk being managed by those best placed to deal with it, and potential wider implications on other activities. The development of more collaborative solutions can also lead to longer-term arrangements than the temporary supply chains that are often used in delivering construction projects today, and can help support smaller businesses in particular, who may be invaluable in the delivery of new solutions but will otherwise struggle to have the space or resources to shift to new ways of working.

In changing the offering a business delivers, it needs to be clear on how best to offer and deliver products and services, bearing in mind the development of new technologies and methods of construction that may work alongside a new solution. It may be that design can largely be automated, or that robots can augment the workforce to install solutions. In considering new models, it is also important to consider how to get from where a business or industry is now to where it needs to be.

Traditional change management processes are unlikely to be suitable to cope with the constantly changing world expected to become reality over the next two decades and beyond. Traditionally, changes to businesses have been gradual, with new technologies or models being added to existing offers until they become financially viable. However, this gradual tweaking of models is unlikely to deliver long-term sustainable solutions. It can be likened to software development; annual upgrade releases tend to offer only marginal differences to the core solution, but occasionally there will be a wholescale rebuild to ensure solutions are fit for purpose at the time of release. The construction industry equivalent is having new entrants to the market offering new forms of construction or new ways of working that can potentially lead to the development of other, complementary services. Changes may challenge a business's dominant logic on how the industry works or the type of business they are, but again they will need to be agile enough to adapt to ever-changing conditions. It will not necessarily be easy, and the cultural change that goes along with technological and business model change should not be underestimated. This is one of the reasons why it is so important to clearly define what a new model will, and will not, involve and deliver.

7.3. Define

In defining what product or service a business will look to offer, it is important to first focus on desired outcomes and understanding whether a business will follow client demand or be a market maker. Taking the step of developing a new market can be a brave choice and lead to facing much more resistance than following demand, but it can also be more rewarding in the long run. Taking advantage of new technologies can potentially smooth the journey somewhat by automating areas that would otherwise require either partnerships with other organisations or investment in further resources, but also by providing additional capability to others in the supply chain. Developing new solutions also needs to take into account which asset types, segments, value chains and geographies are within scope as these will all have an impact on the potential limitations of the offer. For example, what level of customisation will be possible, beneficial and allowed? Who will be the customer, and will systems be installed as part of the service or just sold for others to use? In many ways these decisions are more straightforward for new market entrants as they are not limited by concerns with disturbing existing relationships, but those relationships can help project existing businesses forward if potential partners are in agreement on changing the way they operate.

7.4. Systemise

When looking to systemise delivery or maintenance of assets, the starting point should always be to clearly define the desired outcomes for the system, and then to map out how those outcomes are currently met and by whom. This goes back to Figure 1.1, in that if the wrong things are system-ised then there may not be any improvement in outcomes, and the systems themselves may have a limited market or shelf life. It is important to systemise the right things, not just systemise current processes that may no longer be optimal in delivering desired outcomes. Mapping out existing processes and models using the Five Ps model or similar methodologies helps provide a clearer understanding of areas that can be systemised and supported by other technologies, but it is impor-tant when analysing that it goes further than simply repeating or integrating those processes. This analysis is also useful in understanding potentially new interfaces between the new system and other processes, potential partnership opportunities and also where systems can be used in other sectors with similar requirements.

When looking at different production models, it is important to carefully consider the existing capabilities of the business and supply chain, but not to be limited by them. It is worth looking at the distribution of capabilities across the supply chain – for example, instead of simply locating everything under one roof, there may well be benefits in having a distributed supply chain that reserves some of the delivery to businesses local to development sites. This is likely to have the added benefit of reducing the level of capital investment in production facilities, providing flex-ibility to scale up and down as needed. Considering existing capabilities also enables systems to be developed that are optimised. Not everything necessarily needs to be moved into a production facility if existing capabilities elsewhere in the supply chain more than adequately provide the necessary services to support delivery.

The redistribution of design and production control as a result of systemisation needs to be under-stood. For example, in systemising delivery it is also likely that design responsibility for that system will move to the system provider, with configuration of the system and integration into the wider design being the responsibility of the design team. In developing a system, the additional design responsibility and capability required needs to be understood and in place, and is likely to require different capability than an organisation already has, whether they start from a production, delivery or design perspective; the interface between organisations will change. This is discussed more in Chapters 14 and 17.

Finally, it is worth considering the life cycle of developed systems; how can they be maintained in the future; do they need specialist capabilities or can traditional trades carry out the work? Can they be replaced and reused at the end of their use cycle? These issues may feed back into the sys-tem development process and provide new business opportunities and longer-term relationships directly with clients.

7.5. Automate

Once definition and systemisation have been considered, automation can be looked at as a way of supporting delivery. In doing so, again it is important to look carefully at what can and should be automated, and what is best left as a manual or traditional process. Part of that consideration will be quality and consistency of output along with effort required, but also the availability and type of resource required and whether the right skills are in place to work alongside automation technologies. The entire delivery of an asset will not be automated and, as such, there will always be an interface between automation technologies and humans, and that interface needs to be well managed. When introducing automation, it needs to form a clear part of a defined business model, and not just be applied without due consideration of how value can be created.

The impact of automation on existing and proposed supply chain partnerships needs to be thought through – for example, will services historically provided by a partner be automated and, if so, will that impact relationships and their willingness to work together in other areas that are still required? The introduction of automation is also likely to influence the suitability of contractual models and responsibilities. For example, who takes design responsibility when automated design is implemented?

In addition to the impact on partnerships, the effect on the wellbeing of the existing workforce also needs to be considered. As will be discussed throughout this book, automation should be seen more in terms of augmenting existing capabilities than replacing entire occupations, but the introduction of new technologies can be unsettling for those concerned. Retraining or realigning existing workers needs to be looked at as part of the implementation of automation, as does the impact of new technologies and approaches on the culture of an organisation and its supply chains.

Most forms of automation are reliant on the availability of relevant data. It is therefore critical that data is properly structured, but also that data sources are clear and reliable. In addition, when artificial intelligence is used, it is important to be aware of what data the AI has been trained on, and therefore the suitability of such technologies for a particular application. For example, generative design that has been trained on multistorey residential in Sweden is not likely to be suitable for the design of high-rise residential in the UK. In addition, such technologies need to be trained on solutions that do not use the intellectual property of others without their express permission.

Once desired outcomes have been defined and systemisation and automation have been considered, how delivery can be optimised can be considered.

7.6.   Optimise

With the introduction of new technologies and construction systems there are always likely to be gaps in, or over-resourced, capabilities initially. Whether these imperfections are within an individual business or through the supply chain is in some ways unimportant. What is important is that capabilities can be developed or optimised over time, and that the initial offer should still provide sufficient improvements to be worthwhile from the outset. If new systems are initially more expensive or take more time or more resource, they are unlikely to succeed in the long run unless a clear (and preferably short) path to delivering sustainable improvements over incumbent solutions can be seen.

The development of new interfaces between organisations or processes can cause friction, and to what extent these interfaces can constrain innovation or be dealt with will need careful consideration if new offers are to succeed. Ideally, these new interfaces will not make others' work harder unnecessarily, which is one of the reasons why, when developing new systems, the wider supply chain needs to be considered.

7.7.   Conclusion

When implementing new technologies and systems, the cultural shift is more likely to prove a challenge than the technical shift and, as such, much of the enabling work required needs to focus on the impact of new approaches on the wider supply chain. Many parts of the construction industry are still rather traditional, with some areas having not significantly changed how they operate for several decades. When businesses in those sectors are still making money, potential new solutions need to be able to demonstrate their value quickly to incentivise change and enable trust in alternative approaches to develop. This chapter has provided an overview of some of the considerations when looking to develop and implement new solutions, many of which can be summarised as being clear about what they are looking to achieve, how they will impact others and how they can be incentivised. Chapter 8 describes a number of decisions and considerations that are likely to support systemisation and automation, and minimise abortive work in the future as technologies continue to develop and become available.

emerald PUBLISHING ice

Steve Thompson
ISBN 978-1-83608-599-7
https://doi.org/10.1108/978-1-83608-598-020241010

Chapter 8
No-regret decisions

8.1.   Introduction

As earlier chapters have covered, the construction industry needs to change to become more pro-ductive, more responsive and to deliver more with the same or fewer resources. To enable this to happen, it is likely that there will be a significant shift in the way assets are delivered and operated, and this will undoubtedly involve the application of different technologies, including automa-tion and manufacturing-led construction. However, with the significant breadth of technologies on offer, it can be a real challenge understanding which technologies to apply first and where to start. Following on from the previous chapter which covered some of the things that need to be in place to prepare for the changes that lie ahead, this chapter identifies a number of decisions that can be made, or considerations that are likely to support future development, irrespective of which technologies are applied. The next decade is likely to see an incredibly rapid change in the way that the industry operates and what technologies are used; what follows aims to make the transition a little smoother. The chapter is structured in line with the process steps identified in Figure 1.2, but the recommendations are not ordered in terms of priority or the order in which they should be followed, as these will vary between businesses.

The focus of this chapter is on decisions that can be made by individuals or businesses, not those by Government on industry-wide topics such as regulations.

8.2.   Enable

- *Structure data* – one of the most important enablers for the application of automation technologies is structured data, both product and service data. There is no one structure that works best in all scenarios but, as a starting point, information needs to be digitised and structured so that it can be read by machines and humans. Chapter 10 describes product and service data in more detail.
- *Map data* – where possible, data should be mapped to a common schema, such as the industry foundation classes (IFC), which will enable the data to be exchanged more readily, but will also help in structuring it in the first place and give it more meaning. Mapping data is described in Chapter 10.
- *Develop process models* – these should be used to describe relationships between a business and its suppliers, partners and clients that enable it to deliver products or services. This can form the baseline for the development of new business models and help identify potential areas for improvement or the application of technologies. Simple process models can be used for businesses of any scale and can enable a fresh understanding of what is currently, or what can in the future, be delivered.
- *Develop logical data models* – these are useful in identifying data requirements and relationships and are linked to the development of process models. They enable users to understand how their data may be used and what it will be useful for them to provide. They are explained in Chapter 10.

- *Information management* – consider information management for construction projects, asset management or for products and services, in line with recognised standards such as the ISO 19650 series (BSI, 2019). Information management is covered in Chapter 10.
- *Life cycle thinking* – consider the whole life cycle of a product or asset (not just their initial sale or completion) and how this may be impacted by new solutions or technologies. This can help in identifying new business opportunities such as the leasing of products or providing performance as a service. This is described in more detail in Chapter 21.
- *Education* – the application of new technologies and delivery models will require extensive retraining of the existing workforce, as will preparing new entrants. Digital skills is one of the main areas for development of the current workforce but multidisciplinary skills are also important. Supporting and having links to formal education establishments, such as schools providing relevant T-levels, colleges providing apprenticeships or universities providing professional qualifications, can help in ensuring the next generation of workers have the necessary skills to work alongside the existing workforce in delivering future interventions. Future implications of technologies on work are described in Chapters 22 and 23.
- *Management of intellectual property (IP)* – this is always important but, when applying new technologies that lead to sharing more data, or applying new manufacturing-led solutions, it is important to clearly define what should be protected and what can readily be shared. This should then be reviewed as new solutions are developed or applied. What should be protected in one model may need to be more widely available in another.
- *Connectivity* – the need for, and process of connecting to different stakeholders and models necessary for a business to succeed, needs to be considered early on. Connectivity is described in Chapter 9.
- *Be agile* – the built environment industries are moving into a period where change is likely to become the norm. Under such circumstances, traditional change management practices with a defined start, transition and end state are less likely to be effective. Businesses will need to become more agile to cope with the level and speed of change.

8.3. Define

- *Articulate the voice of the customer* – use models such as quality function deployment (QFD) to identify customer needs when developing new solutions and understand the implications for product development. This is described in Chapters 11 and 12.
- *Focus on desired outcomes* – use tools such as the Value Toolkit to help understand what the customer or society really want or need, ensuring that the right solution is delivered instead of just delivering the wrong solution better. Define a project or initiative outcome profile. This is covered in Chapter 11.
- *Value propositions* – clearly define what the value proposition is: how does an offer meet customer and other stakeholder requirements?
- *The Five Ps Model* – described in Chapter 3, use this to help define what solutions are, what they replace and how they should be compared with alternatives.
- *A portfolio of business models* – where possible, look to offer a number of business models instead of relying on only one or two which may become obsolete with a shift in market conditions or the application of new technologies. For example, there is likely to be a shift towards more work in the refurbishment and repair of existing assets in the next decade, so it may be worth considering developing products or services applicable to that market rather than relying on new-build projects alone.
- *The Five Case Model* – use the principles of the model described in Chapter 11 to help in identifying and business casing new solutions.

8.4.   Systemise

- *Standardisation* – identify opportunities for standardisation and repetition without unnecessarily limiting application. For example, look at developing solutions that enable mass customisation – the efficient production of solutions that enable multiple configurations.
- *Resource efficiency* – look to optimise the use of resources, whether man, machine or partnerships, to deliver a solution at the desired scale and breadth.
- *Life cycle models* – consider the future maintenance, replacement, reuse or recycling of systems, not just their initial application. DFMA on its own is a short-term solution whereas, with careful consideration up front, manufacturing-led construction can provide efficient, adaptable and sustainable solutions and new business models.
- *Integration models* – consider whether it is beneficial to deliver a complete system under one roof or whether it is better to deliver a solution created across different sites and partners to deliver the optimum reach while managing risk.

8.5.   Automate

- *Technology combinations* – identify which technologies make the best combination to deliver the capability to support a business model, instead of duplicating capabilities or selecting technologies that will only be effective in the short term or which rely on the capabilities of others. Chapter 15 describes a number of technologies likely to be used extensively over the next decade and beyond, and Chapters 16–20 include examples of their application.
- *Build for the long term* – plan ahead for the long term, and make decisions on the basis of enabling transition to a future state, instead of selecting technologies based on short-term need alone. The technology timeline in Chapter 15 suggests the likely development periods of different automation technologies.
- *Digitising existing assets* – scan or otherwise create accurate models of existing assets that are to be retained in order to enable manufactured systems to be developed and applied with confidence.
- *Revisit contractual models* – ensure that when new technologies and delivery models are being applied that the commercial model used is still relevant and fit for purpose. Ensure that fee models are appropriate and encourage innovation – for example, where automation technologies reduce design time make sure that businesses are paid based on value delivered, not only on time taken, to make sure businesses are sustainable.

8.6.   Optimise

- *Service life planning* – when considering the use of manufacturing-led solutions, design with the knowledge of the relative life cycles of different solutions, necessary access for future maintenance or replacement and the potential reuse at the end of life.
- *Feedback loops* – using digital twins, collect data on the ongoing condition and performance of products and systems and feed these back into product or system development while enabling complementary systems, such as heating and ventilation, to be optimised.
- *Retain product and material data* – track and maintain product and material data to enable future maintenance and reuse or dismantling at the end of life.

8.7.   Conclusion

The decisions or considerations identified in this chapter will stand readers in good stead for the future application of manufacturing-led and automation technologies in the built environment, and are discussed in more detail throughout the book. It can be noted that the majority of recommendations are technology and solution agnostic, and this is because they provide a base for the adoption of a number of technologies and business types.

REFERENCES

BSI (2019) BS EN ISO 19650: Managing information with building information modelling (BIM). BSI, London, UK.

emerald PUBLISHING   ice

Steve Thompson
ISBN 978-1-83608-599-7
https://doi.org/10.1108/978-1-83608-598-020241011
Emerald Publishing Limited: All rights reserved

Chapter 9
Connectivity

9.1. Introduction

Construction is traditionally seen as a rather siloed industry, whether that be on a project by project basis, by trades and professions, by industry sector, by business or project size or by geography. This structure has limited potential improvements in productivity and output, but has also led to significant mistrust. To be able to overcome the challenges the industry faces over the decades to come, connectivity and transparency need to improve and lessons need to be shared. Because of the traditional organisational structure, in the early to mid-2010s most of the software developed tended to be point solutions – in other words, focused on particular roles or activities with little concern for collaboration. For example, design software was not necessarily developed to enable sharing between disciplines or with contractors. Since the mid-2010s there has been much more focus on collaboration, led by the Government's 2016 BIM mandate (Cabinet Office, 2011) and the interest in better information management. While that has seen significant improvement in how information is shared and used across traditional boundaries, there is still a long way to go before good information management becomes the norm across the entire industry, including among small businesses. While BIM is not as important for small developments as it is for large, complex, multidisciplinary projects, connectivity is just as important (for example, sharing information between designers and contractors, and between contractors and asset owners). Connectivity can involve sharing design information, as-built information, cost data or performance data among other areas and can be a key enabler for many of the solutions discussed throughout this book.

To clarify, there is a difference between being connected and sharing. Sharing means completing activities to a certain stage (such as design) and then issuing it to others for their use. Being connected refers to either a real-time or current state sharing of information to support decision making and can go beyond the sharing of information between software applications.

In addition to the ability to connect between different parties or stages, it is also very important to be able to connect at the right time and speed (which may be real-time collaboration or feedback), but also in an accessible way (for example, connecting design models to present information that will impact on decision making in the right environment). This chapter will look at the benefits and implications of connectivity across sectors, disciplines and asset stages.

9.2. Connectivity across disciplines, stages and sectors

Along with the significant development and broader use of BIM over the last decade has come increased use of common data environments (CDEs), particularly on large or complex projects. CDEs are invaluable in providing centralised access to current information on a project from all sources identified as being relevant, and in ensuring that information to be used is up to date. However, CDEs still rely on users actively looking for information, which is not automatically connected or presented in combination with information from multiple providers, so collaboration can still be limited. Solutions that provide that capability are discussed later in this chapter, but first it is worth looking at how data can be connected, irrespective of project size and complexity.

As mentioned earlier in this section, much of the software developed for the construction industry has historically been focused on point solutions to support specific activities or disciplines and not necessarily how information can be connected. In addition, information was, until recently, only shared in the form of hard copies of drawings and documents, which is still the case on many a small project. As well as design information, there is also construction information that needs to be shared, whether that is to instruct builders to carry out certain activities, to share what has been done or to provide updates on project costs. Solutions to some of these challenges will be covered in later chapters, but a common way of providing connectivity between software is through the use of application programming interfaces (APIs). These enable two or more applications to communicate with each other. For example, Morta uses APIs to connect over 50 construction-specific applications, and provides a no-code environment to enable users to connect and share information with the flexibility of Microsoft Word or Excel with the utility of a database. This enables data to be pulled from, or pushed to, sources such as 3D models and make them accessible for everyone involved in a familiar environment, and can significantly improve transparency. Such technologies can provide organisations with the ability to share information across discipline and project stage boundaries, while avoiding the necessity to learn new, complex software.

Many of the automation solutions discussed in later chapters rely on connectivity behind the scenes to enable them to function. For example, progress sharing technologies that enable on-site activities to be recorded, shared and tracked by all relevant participants need to connect on-site records with the wider platform, and this connectivity enables projects of all sizes to benefit from the latest technologies; connectivity is not only for large projects with large budgets.

Connectivity between project stages and participants is clearly a key enabler for improvements in collaboration, quality and productivity, but it is also worthwhile connecting sectors or even industries beyond construction project boundaries. The Apollo Protocol (IET, 2022) looks at such connections, exploring the potential for and benefits of linking digital twins from the manufacturing, construction and technology industries. If digital twins continue to develop at pace in the construction sector without consideration of potential compatibility with digital twins in manufacturing, then it will be harder to share information between the two worlds. It is important to consider both to ensure data can readily be shared between the two – for example, to enable circular supply chains, provide product performance feedback and to enable new business models such as product leasing. First of all, however, the next section looks at connectivity between design teams.

9.3. Design connectivity

Connectivity between design disciplines and different design software is key to reducing (or hopefully avoiding) clashes, but also enables collaboration and can provide confidence that solutions are likely to be holistic, as input from different disciplines can be seen together. As discussed earlier in this section, there is a difference between connecting and sharing, with connecting being more instantaneous, but also combining and presenting information from different sources in one environment. There are excellent providers of such platforms already in the construction industry such as Revizto and Dalux, both of which enable 2D and 3D data from different sources to be combined into one visual environment.

To enable design information to be shared, processed and developed in multiple applications, and even with bespoke code, the Building and Habitat object Model (BHoM, 2024) has been developed, and is described in Box 9.1.

Box 9.1.   Building and Habitat object Model (BHoM)

The BHoM is a collaborative computational development project that enables multiple point solutions to be connected, along with bespoke scripts developed to solve issues not dealt with by existing software solutions. In simple terms it enables designers to use multiple design tools and bespoke solutions as part of their design development, without losing data through transfer. It includes *adaptors* which allow the exchange of data between bespoke code and external software, and *manipulators* which are the bespoke scripts. A single download provides access to the tool from within multiple existing design tools. Information added to objects makes those objects smarter, which can then be accessed in other software packages. The BHoM is not intended to replace existing software solutions, but to enable better connectivity to make objects smarter and more accessible.

One of the benefits of the BHoM approach is that it can be utilised to develop design solutions and share data during design processes, and so can be used to connect software used by a single author or profession in carrying out their activities, ensuring the most appropriate software can be used while minimising errors in translation and exchange between applications. This differs from tools such as Revizto and Dalux, which are intended to present data from different sources to enable collaboration, not necessarily to aid design development by connecting design tools.

Finally, of course, connecting data for use by different stakeholders and activities will usually require a common language to be used. This does not mean that everything needs to be defined or described in the same way, or even using the same terms necessarily, but that data can be readily exchanged and understood. To this end, Chapter 10 focuses on product and service data and describes both industry foundation classes (IFC) and data dictionaries, which are built for this purpose.

9.4.   Integration connectivity

Connectivity during the construction or integration phase is all about sharing requirements, improving and tracking how products and services are delivered and measuring and reporting on progress. A part of this, however, is communication between different players, whether that is to relay instructions or report on issues. From a supply chain perspective, logistics software and supply chain management platforms (or supply chain digital twins) such as those described in Chapter 18 can provide the required connections, but then further connections are still likely to be required to link in with design or client teams. For connectivity between project teams, visualisation and communication environments such as Sensat (described in Box 9.2) provide shared access in a very visual environment to communicate through. One of the strengths of such an approach is the ease with which they can be understood. Drawings and even 3D models can sometimes be difficult to understand, but displaying them within a real-world context makes this much easier.

Box 9.2. Sensat

Sensat provide a visualisation platform that enables all project data to be presented within a precise visualisation of the real world, including above-ground assets and below-ground services. Such an approach provides real-world context to data and is particularly useful for infrastructure projects covering large areas. The platform makes it easy for all parties to communicate but also includes tools capable of providing accurate volumetric and distance measurements while reducing the need for site visits. The platform works equally well during design, delivery and in use phases.

It is unlikely that there will be one single platform that connects all stakeholders through the integration phase of a project of any scale or complexity, but neither is it necessary. What is important is that those who need to be connected can be, and that connections are easy for the user to make, providing reliable access to the data that they need. Connections for assigning tasks or measuring progress do not necessarily need to rely on the same capabilities as those used to share design or cost data, but they should be able to communicate with those systems. Chapter 19 covers other technologies that can be used to connect different stakeholders and operations during the integration phase, irrespective of project size and scale.

9.5. In use connectivity

During the use phase of an asset, there may well be fewer stakeholders that need to be connected to or communicate with or about the asset. However, connectivity is still important – for example, to enable heating or cooling to be controlled remotely or even automatically, as described in Chapter 20. Connectivity can also support the measurement and reporting of an asset's performance and environmental conditions to identify and manage maintenance tasks or to track the constituent parts of a product or system. All of these are valuable use cases, and should ideally be considered early in the development cycle to enable them to be optimised, while existing assets can also be retrofitted with most of these capabilities.

Digital twins can provide invaluable feedback and interactions with an asset – for example, the National Digital Twin programme's climate resilience demonstrator (CReDo) has infrastructure operators building interconnected digital twins of operational assets, such as power networks and water technologies. However, that level of connectivity is not always necessary and instead can be limited to single use applications, such as remote thermostats or management of maintenance activities.

As new business models develop in the near future, such as the provision of asset performance as a service (for example, with guaranteed energy bills), there is likely to be an increase in connectivity between built assets (or systems) and the provider. This will be to measure performance in use but also to identify when products or systems should be replaced or upgraded. Depending on the product or system, these models are likely to rely on either digital twins or at least connections to enable the exchange of performance and condition data. Without sufficient connectivity capability, these models may well be reactive instead of proactive, or lead to inefficiencies due to the need for either regular inspections or upgrades based on forecasts alone.

9.6.   Conclusion

While earlier chapters of this book have identified some of the challenges the built environment industries face, and the chapters ahead describe many potential solutions, one of the most significant barriers to delivering the changes required is lack of connectivity; connectivity between sectors, between stakeholders (whether those involved in the delivery or operation of an asset or the wider community), between asset life cycle stages and between digital and physical worlds. There are of course technological challenges to providing the necessary connectivity, such as reliable and timely internet access, data security and hardware solutions, power supplies and maintenance. However, connectivity is largely about enabling accurate, reliable and timely communication, and so is just as much of a cultural issue as it is a technological one. It is also something that can be achieved on projects of any scale, and one without a one-size-fits-all solution.

What is important in providing connectivity is that it should be as easy for stakeholders to communicate and interpret incoming information as possible and, with the variety of stakeholders involved in the built environment, this is likely to mean that intermediary languages or technologies will be required to enable translation and exchange. With this in mind, Chapter 10 looks specifically at product and service data and their exchange.

REFERENCES

BHoM (Building and Habitat Object Model) (2024) https://bhom.xyz/ (accessed 30/04/2024).

Cabinet Office (2011) *Government Construction Strategy*. Cabinet. Office, London, UK.

IET (Institution of Engineering and Technology) (2022) *The Apollo Protocol: Unifying Digital Twins Across Sectors*. IET, Hertfordshire, UK.

Steve Thompson
ISBN 978-1-83608-599-7
https://doi.org/10.1108/978-1-83608-598-020241012

Chapter 10
Product and service data

10.1.   Introduction

To start this chapter it is important to first know what is meant by data and information; the two words are often used interchangeably, but they are not the same thing. Put simply, data is raw, unorganised facts such as individual weights, volumes or temperatures. Data can be structured (such as in spreadsheets and databases) or unstructured (such as raw data from sensors). Data on its own do not have meaning; it first needs to be processed or presented in a given context to make it useful. This is information, which can be used to support decision making and understanding.

10.2.   The importance of product data

There is no questioning the important role that construction products and materials play in the delivery and operation of the built environment. Once a construction project has been completed and the delivery teams have left, it is products that remain for their life cycle, or that of the built asset. That has always been the case, but what is likely to be equally important in the future are the data that is supplied and maintained along with those products. As Figure 10.1 highlights, at the centre of manufacturing-led construction, intelligent assets and a circular construction economy lie both products and the data that describe them.

Access to product information of the right quality and at the right time, in a format that can be trusted, is a key enabler of the built environment's digital transformation and optimisation, and will support the application of current and emerging technologies. Awareness of some of these technologies and opportunities has led to a shift over recent years away from BIM to broader information management (IM), a move that means information is becoming even more important to the construction supply chain, including manufacturers. There is no doubt that the move from traditional analogue project processes to the world of BIM has been a significant change, but arguably the change from BIM to better information management reaches wider and deeper, potentially impacting businesses and projects of all sizes. It is in broader information management that manufacturers are most likely to achieve better outcomes for themselves and others, and it is more likely to work across a business rather than BIM, which predominantly deals with how their information is shared externally without direct connection to those providing data. Put simply, accurate and trustworthy product data is to information management what construction products are to a built asset, and so product data is where we start the discussion.

According to KPMG (2021), construction could deliver up to £6 of direct productivity gains for every £1 invested in effective IM, and up to £7.40 in direct cost savings from reductions in delivery programmes, labour time and materials. In addition, cost savings of up to 18% at various life cycle stages can be achieved, so the prize promises to be significant. As already mentioned, the benefits can also spread through the supply chain. For example, a manufacturer providing BIM content does not necessarily benefit from such output other than potentially supporting project delivery. However, if that same manufacturer structures their data effectively as part of the process, and potentially

Figure 10.1 The importance of products and product data (Author's own)

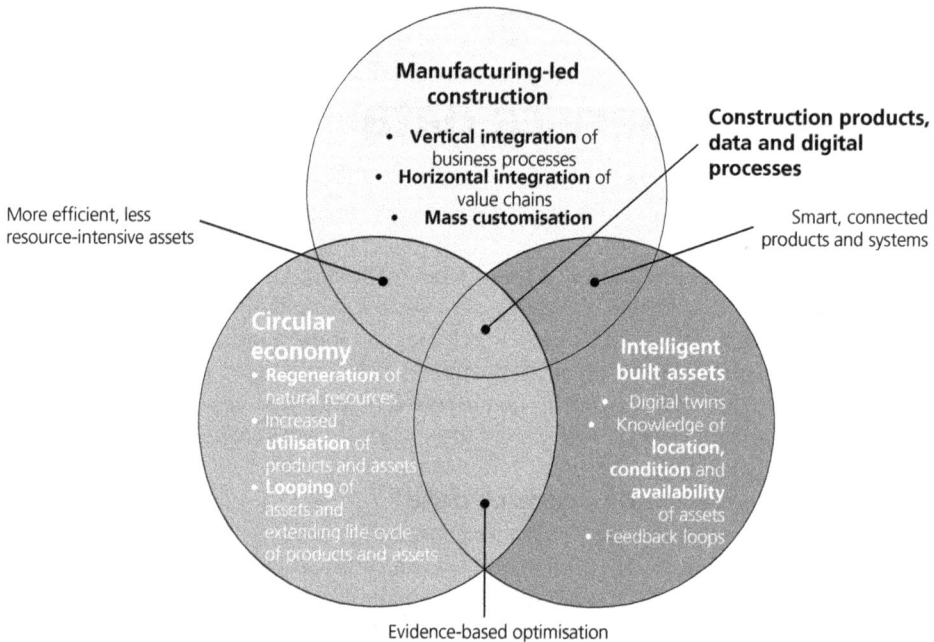

Manufacturing-led construction
- **Vertical integration** of business processes
- **Horizontal integration** of value chains
- **Mass customisation**

Construction products, data and digital processes

More efficient, less resource-intensive assets

Smart, connected products and systems

Circular economy
- Regeneration of natural resources
- Increased utilisation of products and assets
- Looping of assets and extending life cycle of products and assets

Intelligent built assets
- Digital twins
- Knowledge of location, condition and availability of assets
- Feedback loops

Evidence-based optimisation

uses a product information management (PIM) system to manage this data, then they are likely to see significant benefits to their operations as well as providing accurate, up-to-date and reliable data to those that need it. A PIM system can reduce costs in a business, improve data quality and maintenance, enhance customer experience and improve consistency of information across channels, but this requires a lot more than simply providing static BIM objects to clients and specifiers. It requires information to be effectively structured, managed, continuously updated and shared.

There are currently potentially large information gaps within the construction supply chain, a problem illustrated in Figure 10.2. A manufacturer publishes its product information and may share application-specific information with specifiers for a particular project. That information may then be shared by the specifier with the project integrator delivering the project. However, the product itself is typically delivered through a different channel which may include builder's merchants, fabricators, subcontractors and the integrator, all of whom can benefit from better information.

The information gap sits between the information shared with the integrator and the flow of the product. There is the potential for the product to be replaced with an alternative through the delivery process for example, whether that be a minor product variation or a complete change of product, but unless that information is shared with the specifier and integrator, the information supplied may be completely misleading. There is also information that the supply chain may find incredibly valuable which is not passed along with the product, such as performance characteristics that minimise the risk of unsuitable product substitution, or logistics information that needs to be updated. If this information were provided directly from the manufacturer and was accessible along the supply channel, then any actor requiring information would be able to access that information directly. This also prevents the problem of information being lost when passed through an organisation

Figure 10.2 Reducing information gaps within the supply chain (Author's own)

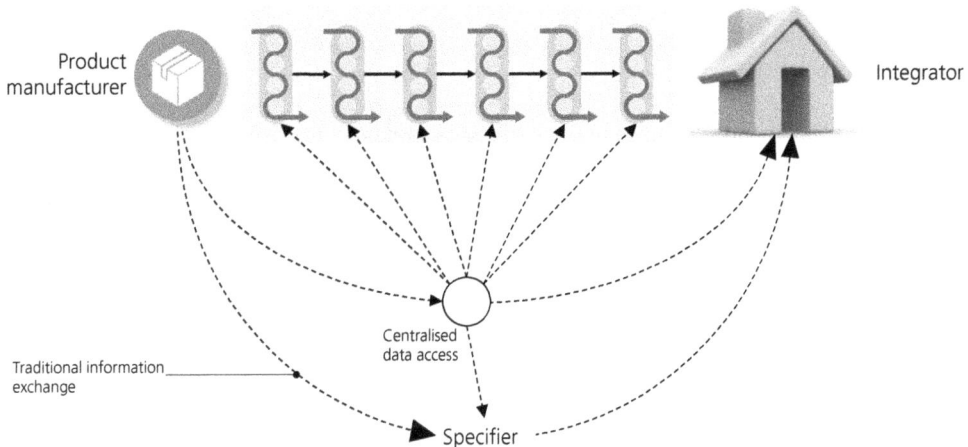

that doesn't appreciate its value to someone further along the chain. Furthermore, there is an ever increasing need in a post-Grenfell world to provide an audit trail for products and related decisions, and this is unlikely to be achieved without a digitally connected supply chain.

It is important to realise that product data do not just refer to data on the final product in use, but also to relevant information on the product as it passes through different states on its way to becoming part of the finished asset, and the processes involved in getting it there. As a result, data needs to be accessible to more businesses, many of which will be small businesses likely to have fewer resources than their larger counterparts. As such, data needs to be accessible through technologies that are ubiquitous, such as mobile devices, not just through expensive hardware and software.

There is little doubt that structured product information will become a key requirement for projects in the future, and what follows here are some of the drivers that give that change real impetus.

10.3. The Building Safety Act 2022, the golden thread and product data

The Building Safety Act 2022 (HMG, 2022) promises to be one of the most significant regulatory changes the construction industry has faced for several decades, driven by the aim to ensure buildings are designed and delivered to remain safe throughout their life cycle. A key aspect of the legislation focuses on how information is shared and managed, particularly product performance and testing information to ensure that the right products are used. From an information management perspective, there is a need to create and maintain a golden thread of information: a digitally stored resource holding information relevant to building safety that needs to be maintained and accessible throughout the life cycle of a building. The golden thread principles (MHCLG, 2021) are

■ accurate and trusted
■ residents feeling secure
■ culture change
■ single source of truth
■ secure

- accountable
- understandable/consistent
- accessible
- longevity/interoperability
- relevant/proportionate.

While the need for structured data is not one of the named headline principles, it is an enabler for several of them. Structuring data enables its accuracy to be clearly determined and for them to be clearly understood and consistent while also being interoperable. For the construction industry to meet the requirements of the new regulatory system, manufacturers will need to structure and share their data effectively, and will be held to account if this is not done, particularly on high risk assets. The data will need to be traceable through the design, delivery and operation of an asset and will need to remain accessible for a number of years after the last of a product has been made available to the market. To achieve that longevity again means that data will need to be structured in such a fashion as to enable exchange through open standards.

In addition to the golden thread, the Building Safety Act 2022 (HMG, 2022) requires information on high-risk buildings to be submitted to the regulator at key gateway points, the first of which is prior to planning permission being granted. This means that product data is likely to be required much earlier in the project life cycle than has traditionally been the case.

Finally, as a result of the Grenfell enquiry, a review of construction product testing and certification was carried out and published in 2023 (Morrell and Day, 2023). This further supported the need for product testing and certification information to be made readily available to those within the supply chain, or at the very least to the regulator. Again, this is likely to drive the need for structured data to evidence the details of a product's testing and certification to enable such data to be machine readable and suitable for exchange.

10.4. Construction digital product and material passports

The Construction Product Regulations across the European Union (EU, 2011) require that all construction products falling under the regulations must also have a Declaration of Performance (DoP). This DoP presents key information on the performance of the product under test conditions. From 2024, any product falling under the Construction Product Regulations in the EU requires a digital product passport. The idea of this passport is to provide all relevant data on a product's performance and characteristics as covered in the relevant standards, and to provide it in a digital format that can be readily exchanged. This is a significant change and move towards sharing structured product data, and it is likely that similar requirements will emerge in the UK. Irrespective, the majority of construction product suppliers operating in the UK also operate internationally, meaning that there will still be a need to provide structured product data which will be available to the UK market. Such data will be required to be accessible directly from the manufacturer, or potentially through a centralised European Commission registry. The goal is to enable industries across different geographies to share construction product data consistently in a standardised way.

Beyond the specifics of construction digital product passports, there is a wider move towards the provision of material passports. The principles are similar, but a material passport focuses on what materials exist in a built asset to enable their potential reuse at the end of their use cycle within that asset. In addition, data in a material passport can have a greater breadth than that of a construction digital product passport, as it does not focus on information relevant to standards alone. A materials passport contains broader information relating to a circular economy in a structured format that can be readily exchanged, but may also incorporate information on the potential future uses of that material, and what may be required to separate the material from others. While called a material passport, it can be at the level of materials, components, products or systems.

10.5. Mapping data sources and requirements through data models

Clearly, construction manufacturing and supply chains are complex, and there are many different data sources and requirements that aid the smooth delivery and operation of assets. When moving towards a more manufacturing-led approach to delivery, it is important to assess the changing data sources and their relevance. The best way to do this is through the development of data models but, before delving into the topic, it is useful to explain some common information management terms used within the manufacturing sector and their relevance for the built environment.

When looking at the benefits of information management to construction product manufacturing, it is important to consider every aspect of a product's or system's development, production, delivery and installation. Figure 10.3 illustrates some of the different information management systems that can be used throughout the product supply chain before any products are even specified, and indicates the relevance to different roles within a business.

Product life cycle management (PLM) covers the development life cycle of products, from initial concept through to market maturity and finally retirement of a product. As such, PLM is typically not directly relevant in sharing data with the construction supply chain.

Product information management (PIM), not to be confused with project information models, are typically centralised systems used to manage product information consistently, both internally and

Figure 10.3 Relationships between types of information management systems for product manufacturers (Author's own)

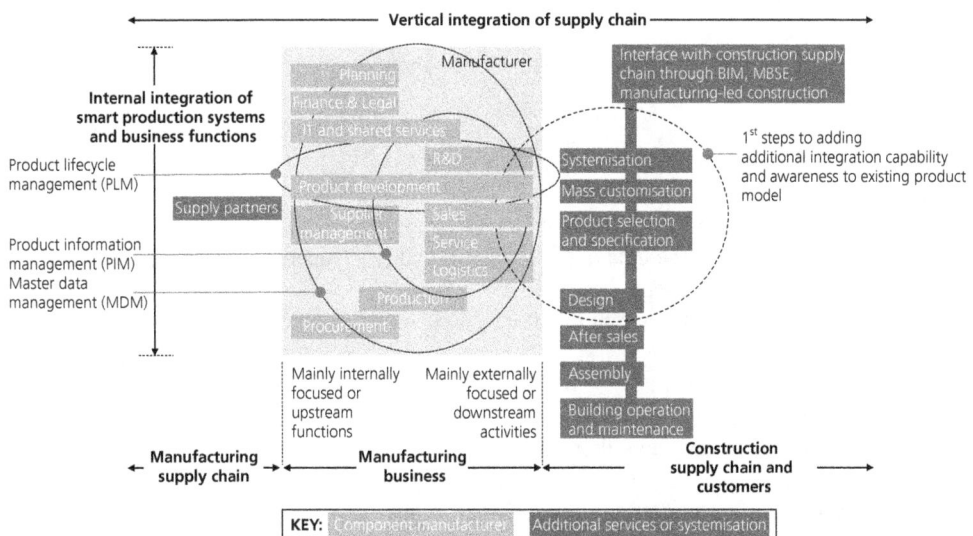

potentially to share with customers and other stakeholders. This is likely to be the most relevant system for the construction supply chain, as it can provide a single source of reliable, up-to-date information on a product, and can provide the link for product data included in a golden thread of information, or in specification or asset management systems.

Master data management (MDM) is management of all data across a business, whether product or nonproduct information. Data is managed in one central process across all relevant territories and can be a significant undertaking. The data managed in a PIM system are a subset of the information in an MDM system.

Depending on the complexity of a business and the range of products covered, a PIM system can be as simple as a spreadsheet or as complex as interlinked databases or proprietary systems. With the application of the golden thread of information as required under the Building Safety Act 2022 (HMG, 2022), it is likely that PIM systems will become even more prevalent in construction product manufacturing as a way of keeping information up-to-date and accessible throughout the life cycle of a built asset. To enable this to happen, however, the first step is to digitise product information across a business or supply chain. This can be a significant undertaking, digitising knowledge from a number of sources. However, it is one that is likely to lead to significant efficiencies (especially with the industry's move to more manufacturing-led and safer delivery practices).

One way to manage the process of digitising the information that lies with product experts is to identify product champions for each product (one product manager may cover multiple products). Product champions are those responsible for managing the input on a particular product, and they do not need to have data management expertise; they are in place for their product expertise alone. Except in the smallest of enterprises, it is unlikely that one individual holds the product knowledge necessary to effectively specify or provide the product data, as one person is unlikely to have sufficient knowledge. As such, the team charged with pulling together data from across a business's products will need to rely on product champions to gain access to the relevant expertise. On top of product data, significant data sits within other functions and, as such, functional champions for relevant parts of a business are also required when considering data for internal use in particular. This process of engaging with experts and understanding the detailed data requirements is often known as discovery.

There are seven key areas that need to be covered when developing information management strategies, and these are

- specification of data requirements
- creation of data
- collection of data and storing it so that it can be accessed
- data assurance – ensuring the data is of sufficient quality to meet specific requirements
- access to relevant data by users
- maintenance of data that is not in active use
- data security.

The discovery phase typically deals with the specification of data requirements, but also understanding what data is produced, how they are managed and where they can be accessed.

Discovery activities typically involve structured interviews across a business (and potentially its supply chain, or even customers), alongside detailed desktop analysis. The discovery does not just cover data sources and requirements, as that would limit the understanding of a business and its needs. Instead, a comprehensive understanding of how the business operates, what standard operating procedures are in place, and how information is used is required. The discovery process therefore unearths the raw information that feeds into the development of data models describing the business and how it wishes to operate. The purpose of discovery is to then make sense of all of these data sources and requirements, along with any proposed new operating models, and to then support the development of data models to describe how data will be managed going forward. Because of the necessarily intrusive nature of a discovery process, a business needs to carefully consider who should carry out such activities. On the one hand, it can be hard for many to cope with the level of questioning from internal colleagues, who may also be new to discoveries and therefore there is a risk of it not achieving the desired results. On the other hand, more experienced but external consultants may also find that businesses are reluctant to share their innermost workings with external teams, so it is a decision that must be fully supported by a business's leadership, and the importance of the activity must be clearly communicated with those that will be involved.

Once the discovery process has been completed and its outputs agreed, the next key task is to develop data models that define how data is to be structured and managed, and the relationships between different entities throughout the scope covered. A data model is a diagram or other form of representation that explains how elements of data relate to each other and to the real world. There are three main types that are relevant here, and ideally each follows on from the previous one. These are

- conceptual data model
- logical data model
- physical data model.

A simple principle of data models is that they should be as simple as possible, but no simpler. In other words, they need to be clear but not simplified to the point where they can cause ambiguity or confusion.

A conceptual data model is a high-level, solution-agnostic description of the real-world information requirements of a business or businesses, and represents the first step in modelling relationships and concepts. It can be used as an initial communication tool to share with key stakeholders to explain the principles of a model – for example, to illustrate how information flows through a process such as product manufacturing. Its role is to get the high level requirements and relationships correct and understood. When looking at managing product data and related data that are relevant to not only internal processes but also the extended construction and manufacturing supply chains, it is recommended that a model be developed covering its whole life cycle. This may seem extreme, given that some of the interactions will not involve the company producing the product, but it will help to identify potential information exchanges and requirements that will need to be managed to take advantage of new legislative requirements or business opportunities – for example, tracking of product data and decisions, leasing of products or systems and future maintenance, replacement or buy-back opportunities.

Once a conceptual data model has been developed and agreed, a logical data model can be developed. Based on the conceptual data model and still solution-agnostic, it provides much more detail on data structures and relationships, typically in the form of relational tables. For all but potentially the largest businesses, it is likely that the development of logical data models will be outsourced to external teams with the necessary expertise.

When developing a data model, it is important to consider future implications to ensure it is as futureproof as possible. Ideally this means that, as requirements change, the model structure remains unchanged but the data within the structure can change to meet the new requirements. This may not always be possible, but it is a good goal to achieve in developing a model.

In the built environment there has been a big shift over the last decade in particular towards object-based modelling. For example, when drawings were created by hand or as line drawings in 2D CAD, information could not be assigned or moved with an object other than through moving notes on a drawing. With object-based models, information can be attached to an object (for example, a door). Furthermore, information from databases can be linked to these objects and kept up-to-date.

The models may consist of a number of information containers or digital repositories. A container can be a 3D model, for example, or a cost plan or a spreadsheet that is project-specific, but the core of the model will be project-agnostic. The use of data for projects should still be included in the model, however, as that is likely to be a key use case for the data managed through the implemented models.

Once the logical data model has been agreed, a physical data model can be developed. It includes the solutions that will be used in operating the model, and so can form the design of a database to manage all of the relevant data. It may be that one logical data model covers several physical data models or systems, and not all are necessarily to be delivered or managed by the same organisation.

In developing the physical data model, decisions will need to be made on whether a manufacturer's product data is stored by the manufacturer or stored and accessed through a third party. Ideally the data will be retained by the manufacturer. This is preferable for a number of reasons, including the traceability of data but also the need to ensure that the data is accurate and current, and that the user can select the data required for their purpose instead of predefining what information should be presented to cover all scenarios. With the Building Safety Act 2022 (HMG, 2022) it is more important than ever to ensure that a manufacturer takes responsibility for managing their product information, and this is much easier to do if they maintain control of it. There is also a risk that third parties charge manufacturers for storing and managing data instead of simply presenting it, but also that they do not fully appreciate the intricacies of the data presented. Around the time of the UK Government's 2016 BIM mandate (NBS, 2015) and in the years running up to that, many manufacturers were charged significant sums to have BIM objects created for their products, even when they were not required. At the same time, many relied on providers to inform them of what information should be included, which led to many manufacturers being unaware of how their products were being described, or the accuracy of the information that was being used to represent them. Clearly today, as then, that is not acceptable and can lead to misleading information on products and their performance, as well as to mistrust.

When developing and implementing data models across a business or supply chain, it is advisable to pilot the models and any business changes on a limited area initially before attempting to change the organisation all at once. In doing so, it is important to identify a part of the business (or, for example, a particular product type) that is representative of the wider business or supply chain but that provides an easily controllable and contained model that does not impact on the business as a whole.

10.6. Structuring product data

When considering how to structure product data in particular, it is important to recognise the difference between data relating to the product and data relating to the application of the product in a defined set of circumstances. For simplicity, the different types of product data throughout its life cycle can be categorised as

- *production data* – data required for the manufacture and distribution of products. This type of data is often overlooked when providing and sharing product data, but can be of significant value for manufacturers and supply chains if managed effectively
- *obtainable product data* – data that describes a product as a thing, and its properties, but which is application- and project-agnostic. Some of this data may historically have been presented in product brochures or technical manuals and may include dimensions, performance and test data
- *product application data* – data on the use of the product in a certain context. This can be split into two
  - *application- or sector-specific data* – for example, for use in a given sector, such as data specifically relevant for use in residential or commercial. This may also include data on how a product performs as part of a built-up system
  - *project-specific data* – data referring to a specific application of a product on a given project. This is particularly relevant for engineered-to-order products or systems
- *installed product data* – data relating to the product in use in the final built asset. This may include warranties and maintenance information.

A manufacturer can provide data relevant to each category, but they cannot provide all data that is required for the application and installed data categories; these need input from other actors and, as such, it is important to identify the source of data provided, along with the suitability of an actor to provide the data. For example, a manufacturer should provide all of the obtainable product data as it all falls within their area of knowledge and responsibility, but it can be dangerous and potentially misleading for them to provide data on specific applications unless they have expertise and possibly test results covering that application.

Before delving further into the type of product data that should be provided, it is important to explain some key terms and concepts relating to data. Attributes of a product or object are known as properties. For example, length, density and material are all properties. To describe a product, these are often grouped into property sets which relate to a specific context. For example, a property set describing a manufacturer may include their name, contact details and location. Each property or property set should have its own globally unique identifier (GUID) to make sure that it is always clear which property or property set is required.

A group of requirements in the form of properties and property sets can be combined into a product data template. These can be industry-agreed information requirements to describe a specific product or set of requirements, or client-specific requirements, but do not include data on a specific product. For example, a template may be developed to provide a consistent way to describe a heat pump, but it will not include data on any specific heat pump or supplier. Once a product data template is completed with data on a specific product, this becomes a product data sheet, and it is this that can be shared by a manufacturer.

Metadata provides the context for the data and is used to describe the properties or property sets. For example, metadata for a property may include property name or property value type (for example, number or Boolean).

A data dictionary is a structured repository of metadata that provides a description of the data. In other words, it helps explain what the data means, and relationships with other data, contexts or objects so that users can understand the data better.

A dictionary includes classes, which are any abstract object (such as a wall or window), concept (such as time) or process (such as installation). Each class provides a description of a set of objects that share the same characteristics. This enables objects of a certain type (such as walls) to have the same property sets that can be filled with the property values of each specific instance. Classes are important as they can be used in combination to provide a detailed set of information requirements for any given construction product type or application, as described later in this section.

There are now internationally recognised standards relating to product data, namely EN ISO 23386 (BSI, 2020a) and EN ISO 23387 (BSI, 2020b). These standards describe how product properties should be created and managed to make them machine readable and interoperable, and also how data templates can be created. However, there are no standards that identify the type of data that should be shared, or how it should be structured other than those relating specifically to the products themselves, which will be covered later in this section. The nearest standard is in the form of ISO 16739-1 (2020c), otherwise known as the industry foundation classes (IFC), which identifies properties and property sets that can be shared and exchanged for a given product or element type. Another useful source of properties to describe products is the electro technical information model (ETIM), but neither ETIM or IFC specifically deal with the different stages in a product's life cycle, such as production and distribution. Instead they tend to describe the product in use. BS EN 17412 (2020d) describes how to define the level of information need based on a set of requirements, but again does not define what data should be included. An example of a framework that can be used to define the level of information and graphical detail required for different purposes is the level of detail guides developed by NBS (2022), which provide examples of what is required for different products and systems as classified in Uniclass.

The most widely used data dictionary for sharing construction information, and a good starting point for managing product data, is the buildingSMART data dictionary (bSDD) (buildingSMART, 2024). It enables users to create their own dictionaries using sets of properties already used or specified in other sources such as IFC and ETIM, or to create their own properties and property sets, but to relate them to certain objects to enable them to be exchanged easily through IFC. The bSDD provides an excellent starting point for anyone looking to define sets of properties to describe construction products or processes and can be used to create properties or property sets to describe any stage of a product's life cycle, not just the application stages of a product. It includes properties and property sets for both IFC and ETIM, as well as properties from any other public dictionary created using or shared in the bSDD.

The first step in defining what product data needs to be shared for a particular product is to clearly identify what class the product falls into. For example, if a product is a window, what type of window is it? It could be classed by material (e.g. wood, aluminium) or type (window, rooflight). For some products there will not always be a straightforward class or category that is suitable, but in these circumstances properties from more than one class can be included. There are different types of product data templates – for example, a product manufacturer may have one template that covers all the data required for a particular product type (the obtainable product information and production information). However, there can also be more specific templates that provide the data relevant for specific use cases. For these templates, data requirements from different classes will also be required, as illustrated in Figure 10.4.

These may include data requirements relating to the specific life cycle stage, the specific application, sector-specific or project-specific requirements. However, ideally these property requirements will be answered where possible from a data set owned and maintained by the manufacturer.

There is no one particular way of defining what product data should be provided and how it should be organised. However, it should avoid duplication wherever possible and must provide clear, unambiguous data. One potential way of organising data to minimise duplication is to use a hierarchical approach such as that described below. One global manufacturer used this approach and reduced the amount of product properties that needed to be managed across their business by over 25%.

Thompson (2016) suggested using the following hierarchy to define data requirements.

1. properties defined as 'essential characteristics' under harmonised European standards
2. properties required in any other national or international standard that is a requirement for the product
3. other industry-recognised requirements
   a. mandated requirements, such as new rules of measurement
   b. non-mandated requirements, such as industry-specific guidelines

Figure 10.4 Product data requirements and answers (Author's own)

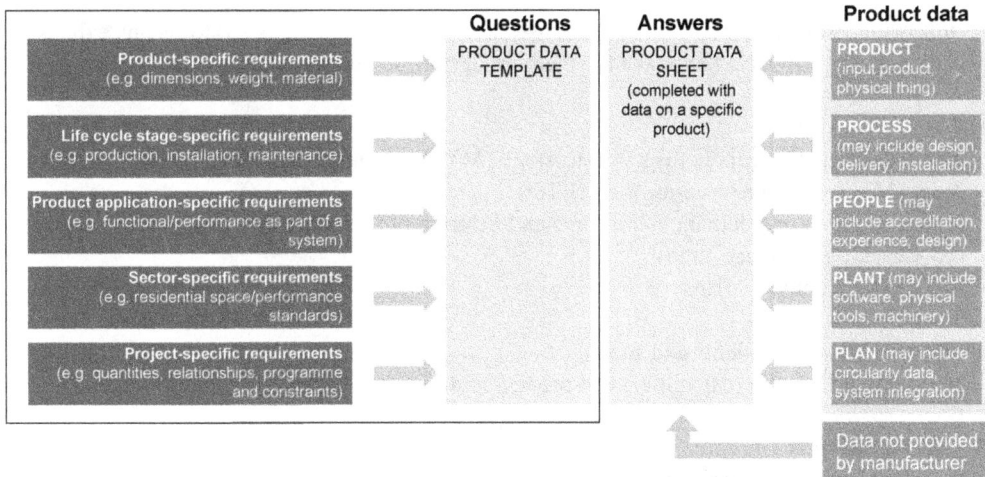

4. industry-agreed requirements, such as properties agreed by industry bodies or relevant authorities
5. user- or client-defined requirements.

Product data can also be organised based on the Five Ps model or, for example, into categories such as physical, chemical, biological and process-related properties. It is always important, however, to consider the potential uses of data, as just providing all available data without a clear understanding of what it will be used for is likely to be unwieldy and confusing for users. Even purely for internal use within a business, it is best to have a clear structure in place. When considering sharing information with stakeholders throughout the supply chain and life cycle, it is essential to be able to map data to a recognised open exchange format, such as IFC. By using the buildingSMART data dictionary or similar approach, properties can be readily exchanged as long as those properties are clearly mapped to relevant IFC objects. It is a common misconception that IFC is just used to exchange 3D models, but far from it. It can readily be used to exchange other data sets, and it provides a very solid base for the exchange of product data outside of 3D models, which means that it can be used throughout the supply chain and can be machine readable.

While IFC does not cover every property that may be necessary to cover the multitude of use cases for product data, there is the capability to create and add custom IFC properties for exchange. To add custom properties, it is strongly recommended that the buildingSMART data dictionary is used, firstly because there may already be a property that is defined that suits the requirements, but also because it means that other users may also then adopt the property in the future.

10.7. Manufacturing and supply chain data requirements

This section discusses some of the types of data that are likely to be relevant to the production and distribution of products. In terms of the framework used in this book and described in Chapter 6, this relates to the life cycle of products and systems, categories A1–A5 (product and assembly stages). It uses the SCOR framework to cover the Plan, Source, Make and Deliver categories. However, in parallel to the production of a product is the specification and procurement process, and so this section also identifies potential use cases for product data in product selection, specification and procurement.

The following list of potential types of and use cases for product data is by no means exhaustive, but is intended to provide a flavour of the plethora of uses of well-structured, reliable product data.

Plan
- identifying whether product is made-to-stock (MTS), engineered-to-stock (ETS), made-to-order (MTO) or engineered-to-order (ETO)
- production planning data including time and location of manufacture
- order backlogs and lead-times
- sourcing and delivery plans
- inventory availability
- supply chain requirements and sources
- product plant, people (capability) and process and product inputs

A1 – Raw materials
- bill of materials (the ingredients), both input products and materials
- hazardous materials data

- supply chain identification and lead times
- inventory availability

A2 and A4 – Transport

- materials handling and inventory data including weight, size and climatic requirements
- loading, unloading and packaging requirements (for example, how products or materials must be stacked and restrained during transport)
- hazardous materials data
- product identification data, such as globally unique identifiers (GUIDs), barcodes and other unique product identifiers
- traceability data to ensure the source of the product and its materials can be traced back and shared onwards. Stock keeping units (SKUs) identify a single product type that can be produced and sold. Two SKUs can be linked to one commercial product, so that building materials produced in two plants to the same specifications can be sold as one commercial product and traced back to source
- declaration of performance (DoP), which identifies key performance data under the EU Construction Product Regulation (CPR)
- merchant's stock numbers and quantities
- load capacity requirements and required plant and driver capabilities
- transportation restrictions including road restrictions
- route planning

A3 – Manufacturing

- required manufacturing process (for example, forming, heating, welding) and suitability for later separation of materials for potential reuse
- plant, people and process requirements
- storage requirements including climate, space and packaging requirements
- materials handling and inventory data
- quality assurance data
- bill of materials
- declaration of performance (DoP), which must be retained by manufacturers for a minimum of 10 years after the product was last made available to the market
- compatibility with other products and materials
- hazardous materials data

A5 – Assembly

- product identification data, such as globally unique identifiers (GUIDs), barcodes and other unique product identifiers.
- product type and product instance data (what the product type is, and when and where the actual product used was manufactured and associated performance and environmental data)
- required plant and people capabilities
- manual handling and installation data
- warranty data
- fixing requirements or recommendations (including to enable future dismantling and replacement or reuse)
- as-built data and certification
- load data (both load capacity and load requirements)
- material passport data

Specification

- product identification data
- lead-time data
- environmental performance (including certification) and environmental performance declaration (EPD)
- bill of materials and manufacturing process to consider embodied carbon and future reuse/separation
- hazardous materials data
- compatibility with other products and materials
- manufacturing location data
- maintenance and replacement data
- test and certification data including that relevant to the required application
- regulatory compliance
- application data and installation instructions
- product constraints (such as suitability for certain markets)
- service life data
- warranty data
- declaration of performance (DoP), which identifies key performance data under the EU Construction Product Regulation (CPR)

Procurement

- product identification data
- declaration of performance (DoP) and other performance data
- environmental performance declaration (EPD)
- assembly/installation location
- identifying whether a product is made-to-stock (MTS), engineered-to-stock (ETS), made-to-order (MTO) or engineered-to-order (ETO)
- specific plant, people and process requirements
- regulatory compliance
- price data
- application-specific use and performance constraints
- maintenance requirements and replacement data
- lead-time data
- packaging, materials handling and logistics requirements
- warranty data
- test and certification data including that relevant to the required application
- packaged quantities
- returns policy and requirements
- excess products management
- quotation, payment, ordering and receipt data
- material passport data.

With all the potential uses of data highlighted in this section, it is clear that a PIM system can be invaluable in managing product data, especially where an organisation delivers a large range of product types. In addition, providing a centralised solution such as that identified in Figure 10.2 will be advantageous in enabling easy access along the supply chain.

10.8. Product data requirements through delivery and asset operation

This section discusses some of the types of data that are likely to be relevant to the installation, operation and end of life of products and systems. In terms of the framework used in this book and described in Chapter 6, this relates to the life cycle of products and systems, categories B and C (use and end of life stages). As with the previous section, the following list of potential use cases is not intended to be exhaustive but to provide an indication of the breadth of uses for reliable product data.

B1 – Use
- product identification data
- product location and access data
- performance data including declaration of performance (DoP) and environmental product declaration (EPD) where applicable
- warranty data
- maintenance requirements and instructions
- material passport data including composition
- testing and compliance data
- manual handling and hazardous materials data
- service life data
- regulatory compliance and restrictions
- traceability of products and materials
- installation dates and installer details

B2, B3 and B5 – Maintenance, repair and refurbishment
- product identification data
- product location and access data
- declaration of performance (DoP) data where applicable
- other performance data and certifications
- maintenance requirements and instructions
- manual handling and hazardous materials data
- service life data
- maintenance competency requirements
- warranty data
- condition data
- regulatory compliance and restrictions
- installation dates
- operational requirements

B4 – Replacement
- product identification data
- product location and access data
- declaration of performance (DoP) and environmental product declaration (EPD) data where applicable
- other performance and certification data
- plant and people capability requirements
- installation and removal requirements

- manual handling and hazardous materials data
- installation dates
- warranty data

B6 and B7 – Operational energy and water use

- product identification data
- declaration of performance (DoP) and environmental product declaration (EPD) data where applicable
- other performance and certification data
- operational requirements
- application-specific performance data
- sensor data on performance in use
- warranty data
- condition data
- service life data

C1 – Deconstruction/demolition

- product identification data
- product location and access data
- declaration of performance (DoP) and environmental product declaration (EPD) data where applicable
- other performance and certification data
- installation and removal instructions and recommendations
- materials passport data
- manual handling and hazardous materials data
- people and plant capability requirements to enable safe removal and potential reuse
- warranty and installation dates
- bill of materials and composition data
- condition data
- installation dates

C2 – Transport

- product location
- materials handling and inventory data including weight, size and climatic requirements
- loading, unloading and packaging requirements (for example, how products or materials must be stacked and restrained during transport)
- hazardous materials data
- product identification data such as globally unique identifiers (GUIDs), barcodes and other unique product identifiers
- required plant and driver capabilities
- transportation restrictions including road restrictions
- route planning

C3 and C4 – Waste processing and disposal

- materials handling and inventory data including weight and size
- hazardous materials data
- declaration of performance (DoP) where relevant and other performance data
- bill of materials and composition data (including fixings)

- condition data
- people and plant capability requirements
- reuse potential.

10.9. Data interoperability

With all the potential use cases for product data, it is clear that many different proprietary software solutions are likely to be used to access or manage the data. It will be completely impractical and even dangerous to provide data in each proprietary format separately, as it will likely mean that information will quickly become out of date. At the same time, if data is only provided in one format and then requires separate input into each proprietary software by users, this can quickly become unmanageable, prone to errors and out of date. The solution is to provide data in a way that can readily be exchanged, and become available in proprietary software solutions from a single source. Such a solution can mean that if the data is updated at source, it can readily be exchanged and updated in all necessary proprietary formats as needed.

To address the need for interoperability, the Government and Industry Interoperability Group (GIIG) produced a report in 2023 dealing with the subject (GIIG, 2023). The report identifies five main principles

- *longevity* – ensuring that data can be findable, accessible and exploitable throughout an asset's or product's life cycle so that it can continue to be valuable
- *security* – ensuring that data is secure and can maintain confidentiality, while sharing non-sensitive data
- *information value* – enhancing the value of data
- *information ownership* – ensuring clear ownership and control of data
- *competition* – providing fair competition between technology providers and users.

Data should be capable of being exchanged between and useable by different software without reliance on the software it was authored in, and manageable in a technology- and service-agnostic fashion. The GIIG report also identifies a number of open standards which should be followed in creating data to ensure it is machine readable and accessible, including standards for geographical data, classification and dates and times.

The longevity of data is key to the future operation and reuse of products and assets; it is important to appreciate that product data has value throughout its life cycle and is not solely relevant to the delivery of an asset. As such, manufacturers in particular may well find themselves being asked to provide data that they are not used to providing. However, it will almost always be information that they already have on their products. Where data relates to specific applications of their products in circumstances that they are not familiar with, it may be that data may also need to come from other sources.

It is crucial that the data provided is reliable, and so should not be provided unless it is known to be correct. For example, when BIM objects were being developed to meet the significant demand in the run up to the UK's 2016 BIM mandate, it was common for objects to be created with properties completed that were not necessarily reliable, and potentially not coming from the manufacturer. This can cause significant risk as users of that data may assume that the data is correct and rely on it for future use. For this reason, it is important to be able to identify the source of any data provided to ensure that it can be relied on and that the source has the necessary capabilities to provide such data. This is a further benefit of providing data directly linked to a manufacturer. In addition, there are clearly other sources of product data that need to provide input throughout the product's life cycle – for example, data of installation, location or condition data and their sources need to be identifiable.

The GIIG report also expresses the need for technology providers to ensure that their technologies can exchange nonproprietary data without loss, amendment, mis-interpretation or additional work for users. To enable this, it recommends that the FAIR principles are followed (Go FAIR, 2023). The principles are

- *findability* – the first step to using data is to find it, whether by humans or machines. To enable this, metadata needs to be machine readable and accessible
- *accessibility* – once the data is found, it needs to be clear how to access it, including any authentication and authorisation
- *interoperability* – the data needs to be integrated with other data and exchangeable between proprietary and non-proprietary systems
- *reusability* – metadata and data need to be well-described so that they can be replicated with certainty.

10.10. Data security and integrity

With the potential sharing of data across multiple organisations covering many use cases over a long period of time, the security and integrity of that data is essential. Data security refers to a collection of measures that keep data from getting corrupted, but it is also important to manage access and permissions for data; data needs to be kept inaccessible to those that may wish to inappropriately take advantage of it or wish to harm it.

When it comes to information security generally, BS EN ISO 27002 (BSI, 2022) provides an internationally recognised framework for setting up an information security management system (ISMS) within a company. Within the built environment, BS EN ISO 19650-5 (BSI, 2020e) defines a security-minded approach to information management. The standard includes an assessment process to identify whether an asset is sensitive, and to initiate a security-minded approach. It identifies an asset, product or service as sensitive if

- there is sufficient risk that it can be used to compromise integrity, safety, security and resilience of an asset, product or service, or its ability to function
- the risk to safety, security and privacy of individuals or communities exceeds the risk appetite of the organisation or supply chain.

It is important to emphasise that the process is not only relevant to sensitive assets but can also be used for sensitive products or supply chains and, where an asset is not recognised as sensitive, a security-minded approach can also be used to protect commercial interests and intellectual property.

To develop a security-minded approach, BS EN ISO 19650-5 (BSI, 2020e) requires

- the creation of a robust governance structure
- the appointment of individuals who will be accountable for the approach adopted
- definition of the individuals responsible for implementing the approach.

The GIIG report (GIIG, 2023) also asks technology providers to ensure that their technologies help asset owners to secure unrestricted ownership and control of their asset-related information, and this should also be the case throughout the supply chain. It also emphasises that while technology must support information exchanges, it does not entitle technology providers to any share of ownership of the information.

Data integrity refers to the safety of data for regulatory compliance and the overall accuracy, completeness and consistency of data, and should be seen as working hand-in-hand with data security. When data is secure, it is more likely to remain complete, accurate and reliable.

There are two types of data integrity

- *physical integrity* – protection of the completeness and accuracy of data as it is stored, maintained and retrieved – for example, when there is a power cut or a disc drive crashes, the physical integrity is likely to be compromised
- *logical integrity* – keeps the data unchanged as it is used in a relational database, and protects from human error or hacking. There are four types of logical integrity
  - o *entity integrity* – ensuring that data isn't listed more than once
  - o *referential integrity* – ensuring that only appropriate changes, additions and deletions of data can occur
  - o *domain integrity* – ensuring the accuracy of each piece of data in a domain (a set of acceptable values)
  - o *user-defined integrity* – the rules and constraints created by the user.

To ensure data integrity, it is important to be able to validate data to check and verify inputs, outputs and transactions. When it comes to product data, a key part of this is ensuring that the data comes from a trusted source, with the relevant competence to make sure it is reliable. The data should not only be of the right type, format and range, but must also be correct. The structuring of data as discussed in this chapter can go a long way to achieving data integrity, but it is essential to consider both integrity and security of data from the outset for them to be achieved; they should not be seen as afterthoughts.

10.11. Conclusion

Well-managed and reliable product data is essential to the smooth delivery and operation of built assets, as well as to the inner workings of product manufacturers and supply chains. The importance of structuring data effectively has never been greater, with the development and application of new technologies, as well as legislation and industry drivers, such as the Building Safety Act 2022 (HMG, 2022) and the Morell and Day Review (Morrell and Day, 2023). However, the structuring of data across a business and supply chain can be a significant undertaking that requires investment in time and money to achieve, and this can often also come with cultural resistance to change. It offers real opportunities for improved efficiencies, however, along with potential business opportunities through the delivery of new support services or opening up of markets by increasing the accessibility of both data and physical products, and by making it easier to specify, deliver and use products. When utilising manufacturing-led approaches, product data management is particularly relevant, as the sharing and exchange of data can be key to supporting new and more efficient ways of working along the supply chain. If new approaches are used that are unfamiliar to those within the supply chain, they should be supported with as much relevant information as possible to make the transition as smooth as possible.

When opening up a company's data to other organisations, it is important to ensure that a manufacturer retains ownership of its data to keep it up to date and reliable, but also secure. When well structured and managed, product data can also play a key role in the future automation of tasks through the specification, delivery and operation of built assets throughout their life cycle, as well as supporting the potential reuse of products at the end of an asset's life cycle. To ensure that data can be readily exchanged and remain accessible throughout its life cycle, it is important to ensure that the data recognises open standards such as IFC, and it is recommended that the buildingSMART Data Dictionary (bsDD) is used to describe or define any properties to enable this to happen.

REFERENCES

BSI (2020a) BS EN ISO 23386: Building information modelling and other digital processes in construction. Methodology to describe, author and maintain properties in interconnected data dictionaries. BSI, London, UK.

BSI (2020b) BS EN ISO 23386: Building information modelling (BIM). Data templates for construction objects used in the life cycle of built assets. Concepts and principles. BSI, London, UK.

BSI (2020c) BS EN ISO 16739-1: Industry foundation classes (IFC) for data sharing in the construction and facility management industries. Data schema. BSI, London, UK.

BSI (2020d) BS EN ISO 23386: Building information modelling. Level of information need. Concepts and principles. BSI, London, UK.

BSI (2020e) BS EN ISO 196505-5: Organization and digitization of information about buildings and civil engineering works, including building information modelling (BIM) Information management using information modelling. Security-minded approach to information management. BSI, London, UK.

BSI (2022) BS EN ISO/IEC 27002: Information security, cybersecurity and privacy protection. Information security controls. BSI, London, UK.

buildingSMART (2024) buildingSMART data dictionary. https://www.buildingsmart.org/users/services/buildingsmart-data-dictionary/ (accessed 19/07/2024).

EU (European Union) (2011) Construction Products Regulation (CPR) No. 305/2011. EU, Brussels, Belgium.

HMG (2022) Building Safety Act 2022. The Stationery Office, London, UK.

GIIG (Government and Industry Interoperability Group) (2023) Delivering valuable data: an interoperability code of practice for technologies in the built and managed environment. https://giig.citizenlab.co/en-GB/folders/delivering-valuable-data (accessed 19/07/2024), UK.

Go FAIR (2023) Fair principles. https://www.go-fair.org/fair-principles/ (accessed 01/12/2023).

KPMG (2021) *The Value of Information Management in the Construction and Infrastructure Sector*. KPMG, London, UK.

MHCLG (Ministry of Housing, Communities and Local Government) (2021) *Building Regulations Advisory Committee: Golden Thread Report*. MHCLG, London, UK.

Morrell P and Day A (2023) *Independent Review of the Construction Product Testing Regime*. Department for Levelling Up, Housing and Communities, London, UK.

NBS (National Building Specification) (2015) https://www.thenbs.com/knowledge/level-of-detail-lod-and-digital-plans-of-work (accessed 31/07/2024).

NBS (2022) https://uniclass.thenbs.com/download (accessed 31/07/2024).

Thompson (2016) *Product Data Definition Document*. Thompson, London, UK.

Section 3

Define

emerald PUBLISHING · ice Publishing

Steve Thompson
ISBN 978-1-83608-599-7
https://doi.org/10.1108/978-1-83608-598-020241014

Chapter 11
Defining the need

11.1. Introduction

As described and illustrated in Chapter 1, before looking at systemising or automating the delivery and management of assets, it is essential to first clearly define what is needed.

Without clear and early definition, there is always the real risk that what is actually needed will not necessarily be delivered. Not only that, but alternative solutions that may have achieved added value in the form of things that really matter may be discarded. This chapter describes different approaches to defining the need for an intervention that can be used as a basis for potential automation and systemisation, potentially across any size of project, programme or portfolio of projects.

It is important to emphasise that defining the need is not the same as defining requirements; defining the need involves identifying and explaining the future state and the gap between that and what already exists, and it comes before the definition of requirements. Requirements identify what outputs must be delivered through an intervention to meet the defined need. While this chapter focuses on defining the need, Chapter 12 then explains how requirements can be clearly defined to enable the delivery of that need.

11.2. Outputs, outcomes, benefits and value

Before moving on to explain their role in moving towards a more value-led approach to definition, it is important to explain some of the key terminology used. As illustrated in Figure 11.1, outputs are the products or services created as the result of an activity – for example, a building delivered through a project. Traditionally, most projects have been defined by output – for example, the built asset at a moment in time. In some cases where the need is clearly understood and well considered, that can be satisfactory, but often a broader and longer-term view can add significant value and more closely meet the needs of the client. As the ICE's *Systems Approach to Infrastructure Delivery* (ICE, 2020) identifies, there is value in thinking about outcomes, not edifices. Such an approach is often described as an outcomes-based or value-driven approach to definition.

Outcomes are the change in state resulting from an intervention and focusing on these instead of over-prescribing a solution can enable the supply chain to bring their significant skills and expertise to the intervention and potentially find new and better ways of achieving the desired outcomes. The Value Toolkit (Construction Innovation Hub, 2023) defines two types of outcome, the first being core outcomes, which are those that are within the project realm and are critical to achieving the project's core mission. The second type of outcome are value outcomes, which fall outside of the project but are equally important. An example would be the desire to create long-term employment opportunities for a local community.

Figure 11.1 Outputs, outcomes, benefits and value (Author's own)

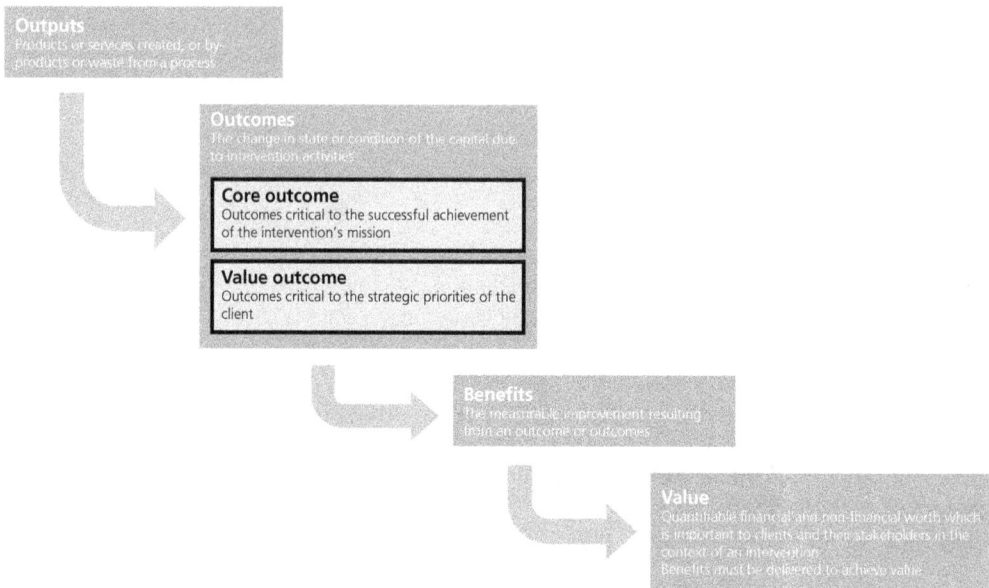

Outputs
Products or services created, or by-products or waste from a process

Outcomes
The change in state or condition of the capital due to intervention activities

Core outcome
Outcomes critical to the successful achievement of the intervention's mission

Value outcome
Outcomes critical to the strategic priorities of the client

Benefits
The measurable improvement resulting from an outcome or outcomes

Value
Quantifiable financial and non-financial worth which is important to clients and their stakeholders in the context of an intervention.
Benefits must be delivered to achieve value

Benefits are the measurable improvements that can result from outcomes – for example, the improvement of exam results being achieved in improved learning environments. Finally, the delivery of benefits achieves value, an improvement in worth (whether financial or nonfinancial – for example, social value). By considering the desired value from the start, the chance of it being achieved is significantly increased. It enables the supply chain to work back from the desired future state (for example, a fully functioning railway or airport) and to ensure that any assets delivered support that future, potentially as part of a wider solution. It also shifts the focus from the initial delivery of an asset to longer-term areas, such as whole life performance and cost, social impact and carbon reduction. Such an approach is commonplace in other industries, such as consumer products, but is relatively new in the built environment. However, the UK Government, among other clients, is moving towards a more balanced approach to definition and assessment of interventions and, while not directly applicable to projects of all types and sizes, the Five Case Model described in the Treasury's Green Book (HMT, 2022) provides a very good starting point on the move to a more holistic consideration of interventions.

11.3. The Green Book and the Five Case Model

The Treasury's Green Book (HMT, 2022) is guidance for central Government on appraisal and evaluation of projects, programmes and policies. While it needs to be read alongside other guidance documentation, it provides useful guidance on approaching how interventions can be considered holistically instead of by cost alone. In doing so, it offers a useful tool that can support the definition of need in the form of the Five Case Model. Other methods to defining need and value described in this chapter can all be used to contribute to the Five Case Model, and so should be seen as aligned to it, not competing against it.

The Five Case Model is the means of developing proposals in a holistic way and is structured around five dimensions which provide different perspectives of the same proposal. These perspectives are illustrated in Figure 11.2, and together provide a broader view of value than is traditionally

Figure 11.2 Five Case Model dimensions (HMT, 2022)

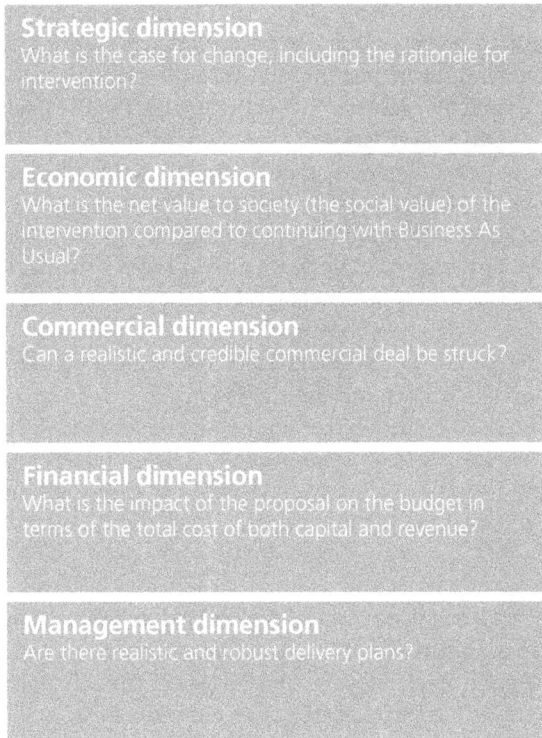

Strategic dimension
What is the case for change, including the rationale for intervention?

Economic dimension
What is the net value to society (the social value) of the intervention compared to continuing with Business As Usual?

Commercial dimension
Can a realistic and credible commercial deal be struck?

Financial dimension
What is the impact of the proposal on the budget in terms of the total cost of both capital and revenue?

Management dimension
Are there realistic and robust delivery plans?

the norm for construction projects. It is important that each of these dimensions are considered together, not in isolation, as they are closely interconnected. Once the value is clearly defined, then and only then can value for money be assessed as the best mix of quality and effectiveness for the least outlay over the life of an asset. The same approach can also be used to assess the value of new delivery models or technologies across a number of projects, such as a new manufacturing-led construction system.

This more holistic approach is particularly relevant to large infrastructure projects where the built assets are there to support other systems, such as trains, tracks and signalling that are required to operate a railway system. The life cycle of different systems and their impact need to be considered; for example, buildings and structures are likely to have a longer life cycle than the technologies used to operate a railway and, as such, need to be capable of enabling upgrades in those systems without requiring significant work to the buildings themselves.

The strategic dimension focuses on the rationale for intervention and includes clearly defining the as-is, or business-as-usual (BAU), to provide a baseline and enable an assessment of the potential value that any change can add. This dimension also includes the definition of SMART objectives (specific, measurable, achievable, realistic and time-limited), which should focus on outcomes to be delivered, not scope alone. The gap between BAU and the future state described by the SMART objectives then represents the need for change.

The economic dimension is where much of the appraisal analysis occurs for public projects and focuses on the potential social value that different options can provide. The *Construction Playbook* (HMG, 2020) identifies that a minimum weighting of 10% should be applied to social value when procuring construction project delivery for public assets to ensure that it is a differentiating factor in evaluation. This clearly emphasises its importance, and tools such as the Value Toolkit (Construction Innovation Hub, 2023) and the National TOMs framework (Social Value Portal, 2023) provide mechanisms to increase awareness and measurement of social value. While social value is likely to be more relevant to public or larger projects, it should certainly still be considered on smaller projects to understand the full value that is required. For example, it may be that the provision of communal green space on a residential development provides important access for a local community to utilise, or the use of local labour can increase employment opportunities. The National TOMs framework described later in this chapter provides an excellent tool to consider and assess potential social value.

The commercial dimension of the Five Case Model looks predominantly at the potential commercial strategies and procurement models that are applicable to deliver the need, and related approaches to risk and cost management. In an ideal world, the commercial model will be developed once the strategic and economic dimensions have been developed but not necessarily finalised (so define the 'what' before the 'how'). However, it is always important to consider the market's capabilities to deliver the need. For private developments this may be partly dictated by the capabilities of an existing supply chain, but that should not restrict the ability to meet a defined need.

The financial dimension takes a holistic view of the net cost to deliver an intervention against the benefits that result. It also considers the affordability of proposals, and clearly that includes cash flow. It should also consider the whole life cost, and the cost of delivering the defined value through other means if the intervention being considered were not taken forward.

Finally, the management dimension looks at the practical implications of delivering the need, whether it can be delivered successfully and monitoring requirements during and after implementation. For built assets that includes the constructability of proposals, but also implications for future maintenance and operation.

The Green Book shouldn't be used in its entirety for all projects: the effort required to do that will at times outweigh the value that it adds. However, using the principles of the Five Case Model can be valuable to projects of any size and type.

The Green Book was clearly not developed for the purposes of defining and assessing built environment assets alone, and so while the principles are relevant, more specific tools are likely to be required to provide more detailed guidance. One such tool, developed by the Construction Innovation Hub and partners specifically for the built environment, is the Value Toolkit.

11.4.   The Value Toolkit

The purpose of the Value Toolkit (Construction Innovation Hub, 2023) is to specify and implement a consistent approach to value-based decision making throughout the life cycle of a built asset. This represents a potentially significant shift from the traditional approach to project definition and solution selection based on the lowest cost to deliver against a design and specification. The first step is to clearly define the client's need in terms of value. What is it that the client actually needs and values? This may not actually require a new, built asset or, if it does, it may not be the built asset initially envisioned by the client or delivery team.

Figure 11.3 The need–value delivery gap (Author's own)

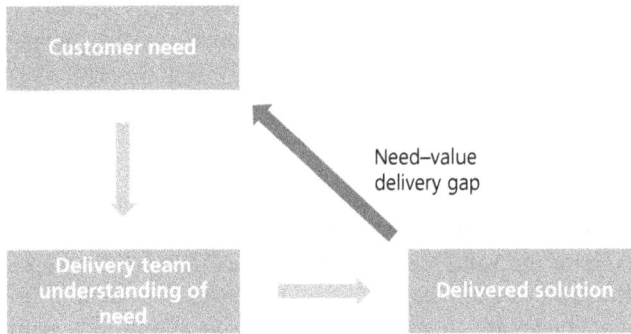

As illustrated in Figure 11.3, there can be a potentially significant gap between what the client needs and what is delivered if the value that the client desires is not clearly understood by the supply chain (including the design team).

To close the need–value gap there needs to be a clear and consistent approach to defining value in such a way that enables it to be measurable, so that it is clear when the necessary value has been realised. To achieve this the Value Toolkit uses the 'four capitals' approach to value definition. It is a recognised approach to defining value that can be delivered by an activity, where a capital is any resource that stores or provides value to people or organisations. As illustrated in Figure 11.4, the four capitals are

- *natural capital* – in the context of the built environment this values the natural environment, and addresses solutions to climate impacts throughout the life cycle of assets
- *human capital* – in the context of the built environment this includes employment opportunities, skills development, health and wellbeing
- *social capital* – in the built environment this refers to influence and consultation, equality and diversity as well as changes to people experience
- *produced capital* – in the built environment this is a combination of capital and operational costs, man-made assets and their efficiency and quality.

Figure 11.4 The four capitals (Capitals Coalition, 2024)

This approach, developed by the Capitals Coalition, helps to assess the potential impact an intervention in the built environment can have on each capital, and forms the basis of the Value Definition Framework (Construction Innovation Hub, 2022), which provides a consistent set of definitions for value categories against each type of capital. This can be used to define an intervention's mission, which has the overall aim of meeting a need, which usually (but not necessarily always) includes building, renovating or maintaining a built asset. The four capitals approach can also be useful in providing a broader approach to risk management than the traditional time, quality and cost model.

Using the Value Toolkit, value can still be defined in a number of ways (for example, social value can still be defined using the National TOMs framework), but the value then needs to be mapped to the defined categories of the Value Definition Framework to provide a consistent way of creating and sharing value profiles and measuring against those. This enables users to demonstrate alignment with national or organisational priorities but also allows different interventions to be compared consistently.

The Value Toolkit has a series of five distinct phases which are illustrated in Figure 11.5. It is important to understand that the value being defined may be delivered through any of these phases, so it not only relates to value that will be delivered once the built asset is in use. For example, employment opportunities may be created through the delivery phase.

The *need* phase exists to define the need and to assess whether an intervention is likely to meet that need. Strategic objectives are defined in the form of core and value outcomes, and these create a reference point for value-based decision making. By the end of this phase, a case needs to be made as to whether a change is necessary or not. The identification of need should remain solution-agnostic and remain clearly focused on the development of a strategic objectives profile, which will be used to evaluate different options during the next phases of the process.

The *optioneering* phase aims to prioritise outcomes and to develop and test potential options against the value profile. Strategic objectives are translated into measurable actions and priorities that can form part of the scope of any intervention. The toolkit comes with a library of metrics which support the definition of measurable outcomes and can be used to validate any proposed solution to ensure it maximises the potential value that can be delivered. At the end of this phase, a preferred option must be identified ready for further development in the design phase.

The *design* phase develops the preferred option in enough detail to enable the delivery organisations to be procured. Development of the solution needs to be assessed against the value profiles created in earlier stages to ensure that there is still the potential to meet value targets. Also at this

Figure 11.5 The Value Toolkit stages (Author's own)

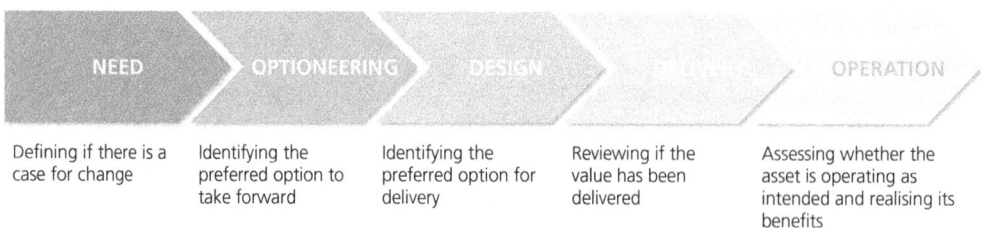

NEED	OPTIONEERING	DESIGN		OPERATION
Defining if there is a case for change	Identifying the preferred option to take forward	Identifying the preferred option for delivery	Reviewing if the value has been delivered	Assessing whether the asset is operating as intended and realising its benefits

stage, the preferred solution is likely to be broken down into deliverable packages of work, with each package having defined outcomes and delivery models. At the end of this phase, the final decision will be to confirm the preferred option for delivery.

The *delivery* phase begins when the contract is awarded and the solution begins to be realised. Performance continues to be measured throughout this phase against the value profile to ensure that it remains on track and to assess whether any further unanticipated value is being delivered. At the end of this phase, the intervention will have been delivered, and the team's performance can be measured against the value scorecard to assess whether delivery has achieved the desired outcomes.

Finally, the *operation* phase covers the use of the asset and continues to monitor whether ongoing value is being realised. The value scorecard updated at the end of the delivery phase provides the baseline for this and any future interventions, including maintenance or remodelling.

Overall, the Value Toolkit provides an excellent framework for defining and measuring value throughout the life cycle of an asset and can be used on any intervention, whether new build or refurbishment, by large organisations or small. At its crudest level, the 17 value categories can be used as a checklist on small projects to prompt further discussions on what value a client is looking to achieve.

Both the Five Case Model and the Value Toolkit highlight the importance of considering social value when defining the need or assessing proposals. A tool commonly used in the public sector to define, quantify and measure social value is the National TOMs framework (Social Value Portal, 2023), developed and managed by the Social Value Portal.

11.5. National TOMs framework

The National TOMs framework (Social Value Portal, 2023) is based around a series of themes, outcomes and measures (TOMs) used to define and measure social value across any intervention or policy. The aim is to provide a link between a broad vision (themes) and strategic objectives (outcomes) which can be expressed as measurable activities (measures). Each measure is quantified using proxy financial measures which provide a considered value for the benefit created by an intervention. The intention of providing proxy financial measures is to provide a common and easily understood metric to compare values of different types; it does not mean that a value can be bought or sold. The five core themes are illustrated in Figure 11.6.

Overall, the framework has in excess of 300 measures of social value, but these are structured in such a fashion that they can be tailored to suit different purposes and levels of complexity. There

Figure 11.6 National TOMs framework themes (Social Value Portal, 2023)

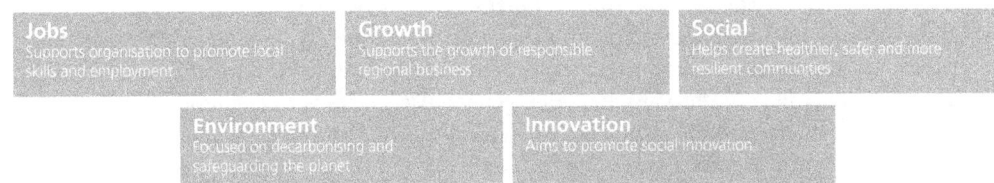

is a light version of the framework that covers 18 different measures of social value, intended for new users or small projects, and a core version which includes 40 measures and is intended to be suitable for most use cases, providing a comprehensive evaluation. The remaining measures enable the most detailed evaluations, including specialist plug-ins for real estate and facilities management. The framework can be used to assess or compare different intervention solutions, to assess the overall impact of a proposed intervention or to quantify the social value already being delivered through an as-is scenario.

As described earlier, the National TOMs can be used alongside the Green Book or the Value Toolkit to measure social value as part of a broader case for change.

11.6. Standardisation across interventions

The need for a value-driven approach is to ensure an intervention is doing the right things, but once that is defined, then and only then does work need to start on defining the need in terms of physical requirements.

There are a number of ways that the construction industry identifies need for projects, and this book is not looking to replace those. Instead, this section explains some of the logic that should be incorporated into the process to enable suitable solutions to emerge and to support both manufacturing-led construction and automation.

If a survey of senior managers in building system manufacturers were to ask what they needed to maintain a sustainable business, it is more than likely that the majority of respondents would mention a clearly visible and realistic pipeline as being in their top asks. To achieve this visibility, however, does not necessarily mean standardising assets to the point that they are identical or even similar in their overall design, but that they may have common space types or performance characteristics, structural requirements and spans or spatial adjacencies. The difficulty that the industry has in identifying such a pipeline, however, is that it is still predominantly working on a project-by-project model, not looking across projects to portfolios or programmes, or across clients. This can then make it almost impossible to identify a consistent pipeline across clients or asset types. Apart from the change in clients across projects, local planning requirements can differ between projects, adding further complexity to the issue.

Working on a project-by-project basis means that it is incredibly difficult to provide consistency in how space requirements are defined or even named, because most projects are still seen as one-offs. However, as the UK's largest construction client, work has been carried out to identify common characteristics across the Government's forward new-build pipeline. This was delivered by the Construction Innovation Hub (2021) and assessed a £50 bn pipeline across five Government departments. It is useful to appreciate that the Government's pipeline includes schools, hospitals, offices and residential (both secure and nonsecure) among other asset types. It is unusual for an assessment to be carried out across such a breadth of asset types, but the results do highlight the benefit of such a move. The analysis suggests that over 70% of the spaces in the pipeline can be met by the same geometrical characteristics (a span of approximately 8 m), and 30% of all spaces are general use spaces (i.e. not part of the specific requirements of a particular asset type), so include circulation, toilets, bathrooms and other general use spaces with common requirements. In many ways, these figures are not a surprise, but understanding that there are common geometrical dimensions across different asset types does not on its own provide opportunities for aggregating demand or delivery efficiencies. There needs also to be consistency in how spaces are named, how they are characterised and how performance requirements are defined (for example, is acoustic performance described in dB or NR?).

The use of digital libraries and tools make it easier to consider repeatability of need across projects and assets, not only because it is easier to reuse definitions, but also because it is easier to virtually test and model different solutions across sectors. As with the naming of spaces, it is important to have a consistent approach to naming of systems and elements to enable reuse of requirements and definitions of what is needed, and for this purpose the latest versions of Uniclass tables provide a solid starting point (NBS, 2022). Using a consistent naming logic means that it is easier to understand the potential pipelines across projects and asset types, but also can speed up the process of accurately defining what is needed. Consistent naming also makes facilities management easier and, as such, the same approach should ideally be used across new and existing assets (or at least existing names mapped to a consistent naming approach so that spaces can be aligned with new). In aligning definition of spaces across assets it can be much easier to understand whether there are enough of a certain type of space (for example, science labs or general teaching spaces in schools). It also enables assessment of whether spaces of a certain type are satisfactory for the purpose for which they are intended. This consistent naming and definition of spaces and assessment of their suitability across existing assets can help in defining requirements for new interventions or assets, in that it provides a clearer picture of what the new or modified asset really needs to provide. It also enables automated rule checking across assets and automated data collection and supports the conversion of requirements to computer-readable language.

Another consideration is the level of definition that is provided to the design or delivery team. If a set of requirements over-constrain a scheme, it can make it harder to provide alternative, innovative solutions to a problem. For example, if the thickness of a wall is specified and standardised, while on the face of it this may seem like a good idea, it can preclude some system providers from offering a competitive solution due to the thickness of their structural systems. On the other hand, standardisation of wall heights and lengths may have little impact on the suppliers of structural systems and their efficiencies as long as the requirements fall within the range that they can provide. It is important therefore to understand the real value of standardising something; does it add value or does it unnecessarily over-constrain potential solutions? This is where early supply chain involvement can be very beneficial at the definition stage to see what potential solutions are out there, where standardisation adds value and where it does not. A good starting point for this when it comes to different building systems is KOPE Market (KOPE, 2023), which provides an overview of many of the potential system suppliers without unnecessarily predefining solutions. It is important to differentiate between early supply chain involvement (ESI) and early contractor involvement (ECI); ESI extends the principle of engaging with Tier 1 organisations to suppliers in Tiers 2 and 3 or beyond who are likely to have an impact on the delivery of a scheme due to their system constraints or capabilities. Engaging at Tier 1 level alone, while still beneficial, does not necessarily provide the level of granularity to really make a difference.

Finally, it is important to consider the level of future adaptability that the client is looking for from a new intervention, which will ideally come out of the definition of required value. It may be that some spaces need to be capable of reconfiguration or shifting to alternative uses in the future to deliver the required value throughout the asset's life cycle, in which case requirements should not be over-constrained based on their initial use alone. This can have a significant impact on the suitability of different building systems, as some are more suited to reconfiguration than others. The definition should also consider the life cycle of different aspects to ensure that future predictable changes can be allowed for (such as the replacement of building services or signalling systems at the end of their life, while avoiding abortive structural works to enable upgrades to happen).

REFERENCES

Capitals Coalition (2024). https://capitalscoalition.org/capitals-approach/ (accessed 20/08/2024).

Construction Innovation Hub (2020) *Value Definition Framework*. Construction Innovation Hub, London, UK.

Construction Innovation Hub (2021) Defining the Need. https://constructioninnovationhub.org.uk/media/vwvji1im/value-toolkit_value-definition-framework_v30-rgb.pdf (accessed 18/07/2024).

Construction Innovation Hub (2023) The Value Toolkit. https://constructioninnovationhub.org.uk/our-projects-and-impact/value-toolkit/ (accessed 16/12/2023).

HMG (Her Majesty's Government) (2020) *The Construction Playbook*. The Stationery Office, London, UK.

HMT (Her Majesty's Treasury) (2022) *The Green Book – Central Government Guidance on Appraisal and Evaluation*. The Stationery Office, London, UK.

ICE (Institution of Civil Engineers) (2020) *A Systems Approach to Infrastructure Delivery*. ICE, London, UK.

KOPE (2023) MMC Market. https://mmc.market/suppliers (accessed 16/12/2023).

NBS (National Building Specification) (2022) https://uniclass.thenbs.com/download (accessed 31/07/2024).

Social Value Portal (2023) National TOMs Framework. https://socialvalueportal.com/solutions/national-toms/ (accessed 16/12/2023).

emerald
PUBLISHING

ice
Publishing

Steve Thompson
ISBN 978-1-83608-599-7
https://doi.org/10.1108/978-1-83608-598-020241015

Chapter 12
Requirements management

12.1.   Introduction

This chapter starts with the assumption that a need has already been defined, and that need includes the delivery, remodelling or refurbishment of a built asset or the development of a new building system or manufacturing-led approach. An asset-focused need and a new solution for multiple interventions can demand different approaches to requirements definition and management, but there are also principles that are relevant to both.

Requirements management involves eliciting real, clear, detailed and measurable requirements from clients and other stakeholders, but also managing those requirements throughout the life cycle of an intervention or solution. Having inadequately defined requirements can lead to scope creep, cost and time overruns or solutions that do not meet the client's real needs. In the construction industry, a potentially significant percentage of project cost and time overruns can be attributed to changes in requirements as projects progress, and many of these changes can be linked to poor requirements definition – in other words, a misunderstanding of what is actually required. Even with the best requirements management systems in place, there is always the possibility that requirements will change or that, as the client more fully appreciates the implications of a project, that requirements become clarified, but requirements management can still better manage changes as they occur to minimise any negative impact.

Learning from other industries such as manufacturing, effective requirements management can lower the cost of development over a solution's life cycle, lead to fewer defects or costly changes, improve traceability of requirements and solutions and aid in the delivery and assessment of value. Many other industries are ahead of the construction sector when it comes to requirements management, possibly because of the unique nature and complexity of built assets and the need for multilayered requirements definition. However, as has already been covered in this book, when stepping back from looking at individual projects, there are significant similarities in requirements across assets that can enable clearer definition. In addition, when considering the development of manufacturing-led solutions across multiple interventions, effective requirements management is necessary to ensure that solutions are suitable for, and deliverable in, their target markets on a repeatable basis. It can be argued that a number of the business failures of manufacturing-led solutions over recent years have been partly due to not understanding the detailed requirements that building systems need to fulfil when applied to or across specific markets, or the requirements of clients and other stakeholders, such as supply chain partners, production facilities and statutory bodies. Requirements management needs to be a collaborative and thorough process, yet it is commonly rushed or not consistently applied within the construction sector. This chapter aims to provide some clarity on the role of requirements management in construction and manufacturing-led solution development, and the potential for it to both use and support automation.

12.2. Levels of definition

Whether defining requirements for a built asset or for a manufacturing-led solution, there is a relatively straightforward hierarchy of requirements that needs to be understood. The elements may have different names, depending on the sector or audience, but the principles remain the same. As illustrated in Figure 11.1, the defined need (outcomes) delivers benefits, which add value. An intervention (whether product or project) delivers outputs, which enable outcomes, but requirements need to be clearly defined to enable the delivery of those outputs. Figure 12.1 illustrates the link between a product or project's vision, its outputs and the different types of requirements that describe what is needed to deliver those outputs.

For simplicity, Figure 12.1 indicates that the relationship ends with a set of functional or nonfunctional requirements. In reality, there are likely to be many different levels of requirements – for example, the requirement to achieve a particular flow rate through a pipeline will lead to a child requirement for a pipe of a certain combination of diameter and strength, and another requirement to have turns of a certain radii.

The hierarchy of requirements outlined in this section enable all requirements to be traceable to a client need and to each other through what is effectively a tree of requirements. For every requirement, there should be a parent requirement above it in the hierarchy, and potentially a child requirement below it. The same logic applies to projects or developments of all sizes, but clearly the more complex the intervention, the larger the requirements tree.

Figure 12.1 Requirements relationships (Author's own)

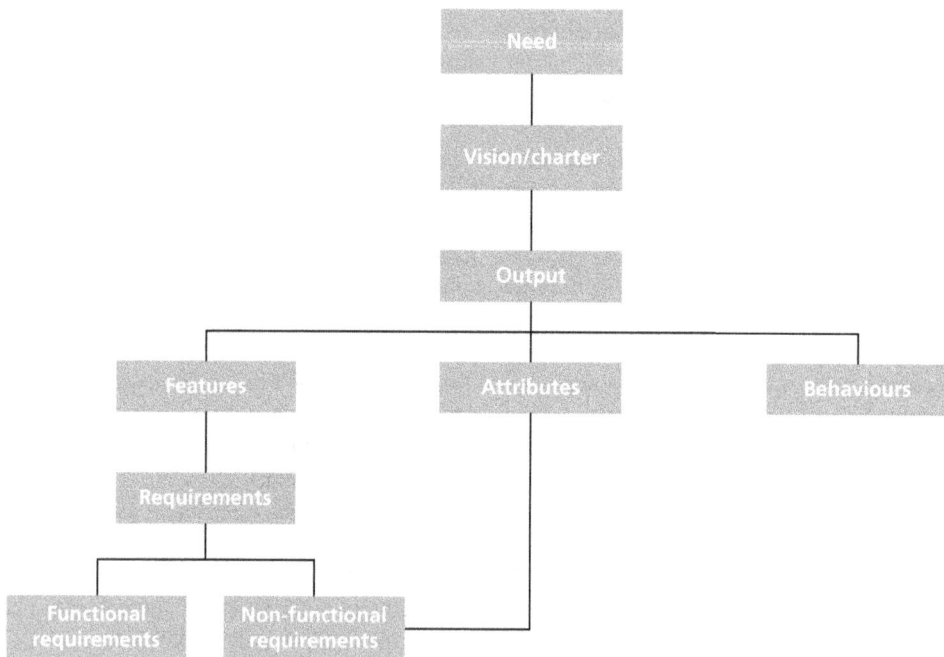

Features are capabilities or characteristics that directly fulfil a need. They are distinctive high-level requirements – for example, the need for an asset or solution to be energy efficient.

Requirements are more granular than features, and typically have a child–parent relationship with a feature or other requirement. Requirements can be split into functional and nonfunctional requirements.

- *Functional requirements* are those that describe what a solution must do in order to fulfil a need – for example, a wall must achieve a U-value no greater than $0.18W/m^2K$ to achieve the feature of being energy efficient. In turn, a requirement to achieve that U-value may be that the insulation have an R-value of no less than $3.2m^2K/W$, and another that it must be supplied by a business within a 50 mile radius of the site. Functional requirements typically play a role in defining the scope of a product or project.
- *Nonfunctional requirements* relate more to the quality of a solution, such as how an asset should look and feel. In other words, functional requirements describe what a solution needs to achieve, and nonfunctional requirements describe what a solution should be.

Requirements are ideally developed top down from a defined customer need, but that is not always the case. For example, there are likely to be specific statutory requirements that relate to certain solutions, such as structural steel or highways specifications that need to be met but that do not come from a project requirement, rather from a solution that has been selected at a later stage in the process.

Attributes are the qualities that a solution should have – for example, durability or comfort – and are closely linked to nonfunctional requirements. They are typically much harder to define and measure than features or functional requirements.

Finally, *behaviours* characterise how a solution acts under given circumstances – for example, lights automatically turn out when the last occupant has left a room.

An important consideration when defining requirements is that they should be as detailed as needed, but no more. In other words, they should be of sufficient detail to avoid ambiguity or confusion, but not so detailed that they provide unnecessary constraints where otherwise there would be none. It is also important to understand the transition required to get from the current state to one where the necessary outcomes have been achieved. Defining a restrictive future state can lead to real challenges in the delivery phase as requirements change, but an effective requirements and change management approach can smooth the transition. Throughout the delivery process requirements may change, but ideally features will remain constant.

As is the case when defining data requirements, it is also important to ensure that requirements are specific. For example, if a specification asks for a pipe that is red, circular and made of steel, that is actually three requirements, not one and so should be separated accordingly.

The scope of a project or product development identifies all the required features and attributes that a solution needs to provide, and the work that is necessary to deliver the required outputs. In language that is more familiar to the construction industry, a performance specification identifies all required features, attributes and behaviours without specifying detailed requirements, and a prescriptive specification is then used to define a solution to meet those requirements. The choice of whether to use a performance (open) or prescriptive specification clearly depends on a number of factors, including the type and complexity of the project or product requirement and the approach to risk.

The next section will highlight some of the differences between defining requirements for projects against those for the development of manufacturing-led solutions, and how not understanding the difference can lead to failure.

12.3. Project or product requirements

In exploring requirements management in the construction industry it is important to emphasise that there are many clients and many providers on any given project. In procurement terms these may be referred to as buyers and sellers, and they exist up and down the supply chain, as illustrated in Figure 12.2. Anyone who buys something is a buyer, and anyone who sells something (whether a product or service) is a seller. The buyer has a requirement (or set of requirements), and the seller then offers a solution to meet that requirement. It is important to appreciate that this type of relationship exists all the way through the supply chain, because they are often overlooked below the asset client and Tier 1 level, yet they occur on projects of any scale and provide a real opportunity to improve the performance of the construction industry.

As Figure 12.2 illustrates, there are requirements that are buyer- or project-specific, but also those that can be applied across multiple projects or interventions (for example, in understanding the

Figure 12.2 Requirement and offer relationships (Author's own)

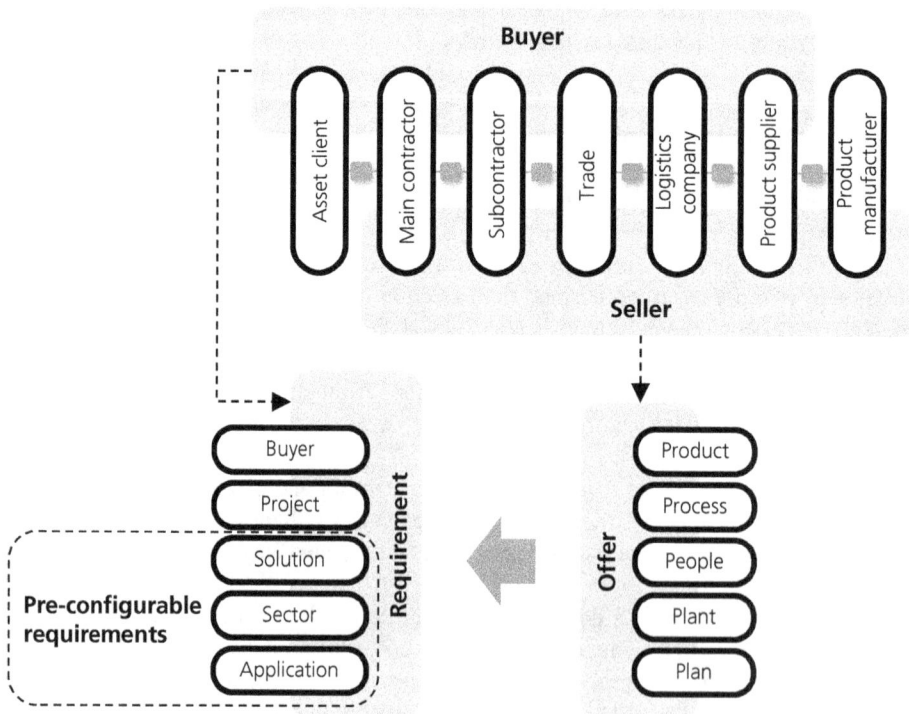

requirements that a new manufacturing-led solution needs to meet to address a defined target market). Firstly, a buyer is likely to have their own project-agnostic requirements, such as payment terms or procurement strategies. In addition there will always be project-specific requirements which are likely to include quantities, programme and storage requirements, climate and project cashflow, or statutory constraints such as planning requirements. These are some of the constraints that are often overlooked when developing manufacturing-led solutions; there will always be planning constraints and a reluctance to pay significant sums upfront of receiving income (for example, paying for volumetric units prior to site installation) and, while these issues may be project-specific, an acceptable range of options can be defined when developing new project-agnostic solutions. If these areas are not considered, it is likely to be difficult to develop and manage a realistic forward pipeline of opportunities as it means each project will have requirements that have not been planned for, or solution providers may target projects that are not realistically deliverable or profitable with their solution in order to keep production facilities running.

As shown in Figure 12.2, there are several types of requirements that can be repeated across projects, portfolios or programmes from a client perspective, or across projects and markets for a solution provider. These can be seen as configurable, as they can be easily foreseen within certain boundaries, and these boundaries can be used to define potential scope and fit for solutions under development. The first of these are solution-specific requirements, which do not necessarily relate to an individual proprietary solution but to a range of solution types. For example, requirements for a structural system may have certain connection requirements, finish types and common dimensional constraints. In addition there may be certain skills that are required to install a solution, such as steelwork erection capabilities.

The second category of configurable requirements is sector-specific, where projects or sector markets have their own regulations or constraints. These may include space standards (which impact on structural spans and performance criteria) or programme constraints, such as possessions on infrastructure projects that mean installations need to be carried out within as short a timescale as possible to avoid disruption and minimise costs.

Finally, there are application-specific requirements. Again, these can be preconfigured as they are likely to repeat across a number of projects and solutions. As an example, there may be specific fixing constraints when a solution is used in a certain context, such as connection details when an internal wall is load-bearing or finishes and insulation where it is fire-separating.

These configurable requirements can form the basis for defining the potential market and scope of new manufacturing-led solutions; in fact, the author has used this approach to develop several solutions in the past. To define solution-specific requirements, the Uniclass Systems and Products tables provide a good, common structure to start the process. Similarly, the Uniclass Entities table (NBS, 2022) provides a useful structure when defining sector or asset-specific requirements. Application-specific requirements can be structured around either the BCIS (Building Cost Information Service) (RICS, 2012) standard form of cost analysis or SFCECA (Standard Form of Civil Engineering Cost Analysis) (RICS, 2024) structures. Other systems can of course be used, but common structures such as these make it easier to map and repeat requirements across projects and solutions and to verify that requirements have been met at a later stage of the process. The need for a clear and consistent approach to requirements management when developing or applying manufacturing-led solutions cannot be over-emphasised; without such an approach in place, the wrong fit between requirements and solutions can occur, leading to failure, inefficiencies or reputational damage. It is within the lower tiers of the supply chain that these issues are often felt, as it is there that interface issues can arise and need to be resolved. Section 12.8 describes how software solutions are available to help manage these requirements and automate the development of requirement trees that make requirements traceable, verifiable and capable of being validated.

In addition to the different types of requirements that can occur, the scale of development or application should also be considered. Over recent years the Construction Innovation Hub and partners have developed a quality process called Construction Product Quality Planning (CPQP) (Construction Innovation Hub, 2022b), which is a structured methodology aimed at supporting the development and introduction of new construction products at scale. The process includes methods for defining requirements from a customer need, and is based on approaches common in other industries such as manufacturing. It is a very useful methodology, but is very resource-intensive and so should only ever be used for proposed solutions of significant scale and consistency. For smaller interventions and product developments, some of the tools used within CPQP can be simplified (quality function deployment, for example), and other methodologies can also be used, as described in the next section. Irrespective of scale, it is likely that to define requirements effectively will require collaboration: it is unlikely that one individual or organisation will be able to completely define all requirements for a project or product development without assistance.

12.4. Writing requirements

The first activity when defining requirements for an intervention (whether a construction project or new manufacturing-led solution to multiple projects) is to baseline the as-is situation; without knowing the starting point of a journey it is unlikely that you will find a suitable route to achieve the desired outcomes. When creating a baseline it is important to provide a starting point from which improvements can be measured; without that solid foundation it can be difficult to understand the value to be delivered through an intervention and whether it is worth the investment of time and resources. For example, a requirement stating that a solution must be at least 20% quicker and cheaper than traditional construction is not a measurable requirement, unless traditional construction is very clearly defined first.

Once a baseline has been created, there are a number of key steps to follow in identifying and managing requirements, and these are identified in Figure 12.3 and described in more detail in the following sections.

12.4.1   Identification

As illustrated in Figure 12.1, requirements definition follows the project (or solution) vision or charter, and the features and attributes that flow from that. The vision will primarily refer to the operational phase of an asset, but the features and requirements should refer to the whole life cycle of an asset, product or service, including the delivery, operational and end of life phases.

Figure 12.3 Requirements management steps (Author's own)

For the development of a manufacturing-led solution, areas considered for requirements should include product development, storage and distribution, maintenance and repair, application, reuse and disposal as well as use stages. A new system is likely to mean a change in the way things are done for follow-on trades among others, and so requirements need to be very clear and communicated effectively to ensure that changes from the norm are properly understood by all relevant stakeholders. For example, no matter how precise and efficient a new volumetric construction is, if a follow-on trade such as an electrician or plumber penetrates a construction because access has not been considered or understood, then potential benefits can be wasted. As such, it may be that requirements are defined in more detail than is tradition-ally the case to ensure that the full impact of new methodologies can be understood, and root causes of any difficulties addressed. In this way, the identification of requirements for new technologies can be more closely matched to requirements management in software development than on traditional con-struction projects, and can require thorough interviews with a range of stakeholders to develop the level of granularity required. If requirements are only defined at a high level, it can lead to false assumptions being made on the suitability of new technologies which can result in scope creep or disenfranchise-ment of supply chains who have not been consulted in the process. This is all too common in solution development, and it is these interfaces that can lead to the success or failure of new approaches. For this reason, it is worth spending time exploring all potential stakeholders impacted by new manufacturing-led solutions, and then consulting those stakeholders as part of the requirements discovery process.

12.4.2 Specification

It is important wherever possible to ensure requirements are measurable, so that delivery against them can be confirmed. There should also be no room for interpretation, and one simple way of clarifying a requirement is to ask the question 'What would you consider to be a failure to meet this requirement?' Another approach commonly used is known as the 'Five Why's' method. To get to the real root cause of an issue, one asks on average five questions beginning with 'why'. The following is an example from Toyota (Ohno, 1988) about a machine that stopped working.

1. Why did the machine stop?
   There was an overload and the fuse blew.

2. Why was there an overload?
   The bearing was not sufficiently lubricated.

3. Why was it not lubricated?
   The lubrication pump was not pumping sufficiently.

4. Why was it not pumping sufficiently?
   The shaft of the pump was worn and rattling.

5. Why was the shaft worn out?
   There was no strainer attached and metal scraps got in.

Without repeatedly asking why, managers would simply replace the fuse or pump and the failure would recur. The specific number five is not the point. Rather it is to keep asking until the root cause is reached and eliminated.

When defining requirements, there are certain characteristics that should be included, such as ensuring they are

- specific
- measurable and testable

- clear and concise
- accurate
- human and machine readable
- feasible and realistic
- necessary
- prioritised
- consistent (not contradicting other requirements).

Ensuring that these attributes are considered will greatly assist in providing good requirements.

On complex projects or in the development of new manufacturing-led solutions, the number of requirements can be significant, and without any form of prioritisation it can be incredibly difficult to determine which requirements really need to be addressed first. In some cases this will be clear for all to see but, especially when dealing with detailed requirements, it can prove a real challenge. Different software solutions manage requirements in different ways but, as a guide, a simple methodology is to categorise them as 'must have', 'should have' or 'nice to have'. This logic can be applied to developments of any scale and can be referenced back to stakeholders to validate requirements before solutions are developed.

When preparing requirements management plans manually, it is also useful to categorise them to make it easier to understand the relationship with other requirements and, as suggested in Section 12.3, Uniclass, BCIS and SFCECA provide useful categories for this purpose.

12.4.3 Mapping

Once a requirement has been specified, it also needs to be mapped to parent requirements and back to required features and outcomes. This process is another check that a requirement is both logical and indispensable. If a requirement cannot be traced back, it suggests that either it is not a critical requirement or that further work needs to be done in defining related requirements. Requirements management software can make this process significantly easier, but where it is not used (for example, on small developments), it can be useful to define requirements in a spreadsheet, one requirement per row. Doing this enables requirements to be easily filtered by categories, priorities or sources, for example, to make it easier to understand relationships.

12.4.4 Validation

Verification and validation of requirements is covered in more detail in Section 12.7.

12.4.5 Tracking

Once requirements have been defined and accepted, it is important to track whether solutions have met those requirements, and for that purpose a useful tool is the requirements traceability matrix. A requirements traceability matrix is a report (ideally database) that demonstrates the relationship between requirements and other artefacts, such as evidence of compliance or test results. It is used to prove that requirements have been fulfilled. It is typically included in requirements management software but, again for smaller developments, can be in the form of a spreadsheet. What is important is that it provides a live story of how projects are progressing against defined requirements.

12.4.6   Maintenance

Once defined, requirements need to be maintained throughout a development to ensure that they are still relevant, and requirements need to be added, edited or removed as necessary through the process. A requirements management plan is typically used to define the process that will be followed to maintain requirements, including change processes that need to be followed for any changes once the initial requirements have been baselined. Again, the easiest way to present a requirements management plan is in tabular form, where all features and related requirements are listed in one table, indicating status and relationships.

The next two sections of this chapter outline two approaches to requirements management that are widely used in other industries, and are particularly relevant to manufacturing-led approaches.

12.5.   Systems engineering

Systems engineering and model-based systems engineering (MBSE) are described in Sections 5.3 and 5.4 of this book, respectively, and can be invaluable in managing requirements from definition through to fulfilment. While requirements management concentrates on capturing and managing the specific requirements and constraints of a development, systems engineering focuses on the overall system design, integration and delivery. As with other approaches mentioned in this chapter, systems engineering can require significant investment in time, money and resources, but the principles of the systems 'V' illustrated in Figure 5.3 can be used on developments of any scale, ensuring that requirements are clearly defined and linked, and that delivery is tested against those requirements at appropriate levels of granularity.

12.6.   Quality function deployment (QFD)

While there is nothing stopping it being used for the development of requirements for a particular construction project, the quality function deployment (QFD) methodology used in many product industries is particularly relevant for requirements definition for new product development. It provides a structured methodology to capture the voice of the customer and translate these into technical specifications and can be used on projects or initiatives of any scale. The basic tool used in QFD is known as the 'House of Quality' and is illustrated in Figure 12.4.

The Construction Innovation Hub (2022a) produced a useful guide to QFD for use in the development of new construction products, but the principles of QFD are quite straightforward. The purpose of QFD is to capture and prioritise requirements based on their importance to the customer. Starting with the 'whats', key customer requirements are identified and included to the left of the house structure. As an example, these can include budgets, space requirements, or the desire to be energy efficient. The technical requirements from the development team (for example, architects, engineers and production teams) are listed in the 'hows' section. These may include performance of elements, regulations or the constraint to use local suppliers. The relationships between the whats and the hows are now assessed to identify how strongly the hows affect the ability to achieve the whats. These assessments are then added up in the technical assessment to identify the relative importance of each how.

In the roof of the house, the interrelationships of different hows are identified, which gives an indication of how the fulfilment of one technical requirement may impact on another. Finally, an analysis of the ability of competitors to fulfil customer requirements is undertaken on the right of the house, which can then feed into solution development to differentiate from the competition.

Figure 12.4 House of quality (Author's own)

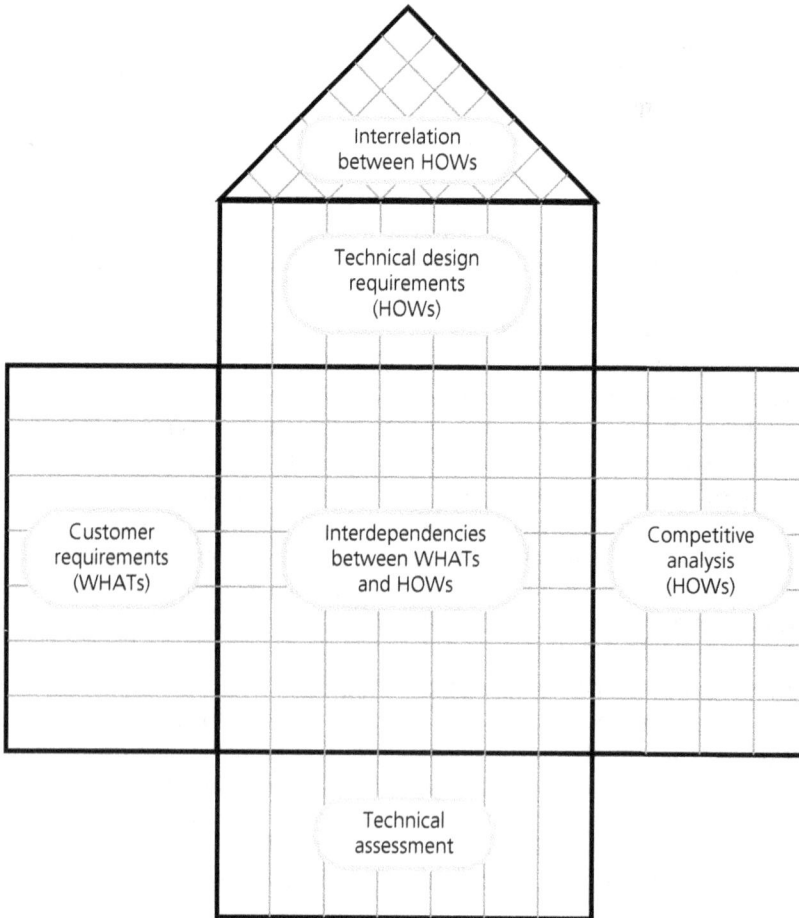

12.7. Verification and validation of requirements

Requirements validation is the process of checking and confirming that the requirements defined accurately capture the needs and expectations of the identified stakeholders, and that ambiguities and conflicts in requirements are resolved prior to development. In other words, validation is a check that the right requirements have been recorded. Verification on the other hand is checking that the requirements have been recorded correctly. Verification is relatively straightforward using requirements management software, whereas validation usually entails going back to requirement sources and checking that the requirements accurately describe what is actually needed.

12.8. Requirements management software solutions

In reading through this chapter, you may think that all this requirements management at different levels of hierarchy seems unnecessary and cumbersome. In a traditional, analogue model it may well be cumbersome, but today thankfully the software exists to make it manageable and to deliver the required benefits. In an analogue world, detailed requirements may be described in a requirements management plan document, but software can manage requirements digitally, which means they can be treated as being live, accessible and always up to date. The principles of a requirements management plan still apply although, in that it describes how requirements will be analysed, documented and managed throughout a project or solution development.

One of the main benefits of requirements management software is the ability to access and collaborate live on requirements from anywhere through a web browser, and to be able to trace detailed requirements back to high-level features, creating a clear digital thread. Most tools also provide access controls so that access can be limited to those areas relevant to a particular stakeholder, instead of overwhelming users and preventing unauthorised changes.

Software tools also support change management processes to add, remove or edit requirements, but also to provide updates where requirements have been fulfilled, with supporting evidence. Tools such as Jama Connect, IBM Engineering Requirements Management DOORS also have the ability to reuse requirements across different initiatives, which is ideal where preconfigurable requirements exist.

It is difficult to automate the process of requirements identification and specification, particularly on individual projects, as it requires elicitation of knowledge from a range of stakeholders, and then validation that those requirements are the right ones. However, where preconfigurable requirements exist (for example, when applying manufacturing-led solutions to multiple projects), then automation is possible, and can quickly provide certainty. The author has previously developed proprietary tools for this very purpose, where detailed location and sector-specific requirements were automatically configured for a building system to suit specific project requirements, with only client- and project-specific requirements needing to be added.

Once requirements have been drafted, artificial intelligence can be used to verify that the requirements are of sufficient quality – for example, IBM's Engineering Requirements Quality Assistant uses Watson AI for this purpose.

12.9. Conclusion

Requirements management has always been important in the construction industry, but arguably never as important as it is today. As described in earlier chapters, the industry needs to find new and more efficient ways of delivering and maintaining the built environment, and the first step in achieving this is to clearly define what is required. Lessons can be learned from other industries, such as manufacturing and software development, both of which have used requirements management in different forms for decades, and software tools have been developed specifically for the purpose. Just as important is the use of requirements management in developing and applying new manufacturing-led models for application on multiple projects, to ensure that there is a good fit between the developed solutions and the markets that they will operate in. This chapter has provided some context to requirements management and identified some methodologies and techniques that can be applied to construction projects or product developments. Effective requirements management provides a good foundation for automation throughout the design, delivery and operation of assets, and the ability to constantly check that requirements are being met.

REFERENCES

Construction Innovation Hub (2022a) *Quality Function Deployment Guideline*. Construction Innovation Hub, London, UK.

Construction Innovation Hub (2022b) Construction Product Quality Planning (CPQP). https://constructioninnovationhub.org.uk/our-projects-and-impact/platform-programme/assurance/cpqp-quality/ (accessed 31/07/2024).

NBS (National Building Specification) (2022) https://uniclass.thenbs.com/download (accessed 31/07/2024).

Ohno T (1988) *Toyota Production System – Beyond Large-Scale Production*. Productivity Press, New York, NY, USA.

RICS (Royal Institution of Chartered Surveyors) (2012) *Elemental Standard Form of Cost Analysis: Principles, Instructions, Elements and Definitions*, 4th (NRM) edn. RICS, London, UK.

RICS (2024) Classification of costs. https://www.isurv.com/info/1515/benchmarking_infrastructure_costs/11540/classification_of_costs/4 (accessed 31/07/2024).

Steve Thompson
ISBN 978-1-83608-599-7
https://doi.org/10.1108/978-1-83608-598-020241016

Chapter 13
Product platforms

13.1. Introduction

Product platforms are nothing new in the world of manufacturing and most product sectors; in fact, they have been around in different guises for several decades. For example, the Apple iPhone is based on a product platform that allows for the creation of multiple models and variations, while still sharing many common components and features. Product platforms are, however, relatively new to the construction industry. There are a number of reasons for this, and none greater than the traditional complexity of the industry, the typical uniqueness of every 'product' (building) and the distance that traditionally exists between manufacturing and the client. So, what are product platforms, and why and how are they used in other industries?

A product platform is essentially a collection of assets that are shared by a number of products. Those assets are not necessarily shared physical products but can also be processes, plant, planning (knowledge) or people, which goes back to the Five Ps model described in Chapter 3. Here, products can be physical components or assemblies used across multiple projects or assets. Processes are repeatable activities used to deliver consistently, such as design processes or assembly. Plant is common equipment or production capabilities (which may be provided at a number of locations to an agreed specification). Planning is the design and technical know-how to deliver products (which may include licenced designs and technical specifications, or assembly techniques) and people are the common team, networks or culture used to deliver. In many ways, the mix of asset types used does not matter; what matters is the quality and suitability of output that a platform approach can deliver when carefully scoped, developed and managed. However, using common components or assemblies alone across multiple instances does not constitute a platform. A platform approach should look holistically at delivery of common assets and how improvements in efficiency and productivity can be achieved and repeated across multiple projects, whether that work be carried out on-site, off-site or whether activities are even necessary at all.

Interest in platforms for construction is increasing as the potential to link manufacturing efficiencies with clients is becoming more feasible with the rapid progress of technology and digitalisation across the industry. It is leading to greater understanding of how mass customisation can deliver those efficiencies at the same time as responding to unique project requirements, and how products and processes can be shared and repeated across instances.

Mosca *et al.* (2020) identify three types of product platform, each relevant to construction in different ways. These are

- *scalable* – a core product is used, but variations in one or more parameters produce different results or capabilities. An example of this type of platform is the Black & Decker drill motor platform. This platform could be scaled along its length to generate different power outputs for different markets

▦ *modular* – core features or components remain constant, and peripheral modules or assemblies are added utilising standardised interfaces to provide customisation. This is the platform type that is likely to be most commonly used in construction (possibly in combination with scalable platforms) and covers platforms developed in line with the Platform Rulebook (Construction Innovation Hub, 2023a)

▦ *generational* – products are delivered that can enable changes as future generations are developed. An example is where a product can be upgraded by replacing a component or assembly using standard interfaces to provide a product with superior performance. This type of platform is unique in that it focuses on enabling changes over time, and so in construction terms may consider future reconfigurations or upgrades throughout the life cycle of an asset.

It is important to note that a platform does not need to be delivered by one organisation alone; in fact, it may be more likely to succeed if a network of partners is utilised to deliver the necessary capability to a well-defined plan, examples being the Seismic Platform® described in Box 13.1, or Modulous in Box 14.2. In both instances the technical know-how, intellectual property and supply chain management sits with the platform provider; however, in both instances the client can either be an integrator (or main contractor in old language), or the client or developer directly.

Box 13.1   The Seismic Platform®

Figure 13.1 Patented Seismic connection detail (Source: Seismic Group)

Seismic Group and partners have developed a standardised corner connector that enables modules and panels from different suppliers to be connected together both horizontally and vertically. Illustrated in Figure 13.1, the connector is patented and can be licensed for use. This means that more than one supplier can be used to deliver a project without the need for significant change being required to change building system. It can significantly reduce the risk of one supplier no longer having the capability to deliver a solution. The size of modules and panel dimensions are not standardised, providing flexibility in application, which is supported by common solutions for floor, wall and ceiling cassettes and building envelope.

The platform is suitable for use across multiple sectors, including education, commercial and residential. The platform promises to achieve 70% faster assembly and 70% less carbon than traditional construction. Due to the design of the kit of parts and standard connector, the system can be disassembled and components reused at a later date.

Where a platform is developed around a core offering that is in effect a back end to what the customer buys, it may be that multiple organisations configure the core offering and add to it to provide a number of products tailored to different consumers. This has the effect of separating the platform provider from the end customer. The alternative approach is to provide a fuller offering that more directly interfaces with the customer, while still providing the same level of customisation by configuring the core and peripheral elements. Both approaches are relevant to the construction industry, and the choice for a platform provider is likely to be based on existing capabilities and experience within a given market. One area that is crucial to both approaches, however, is knowledge of customer needs and priorities, as well as an understanding of how a platform approach can help meet those more effectively and efficiently than other approaches, such as standardisation or off-site manufacture alone.

13.2.   Product platforms in the built environment

Over recent years there has been a significant increase in both awareness and interest in product platforms in construction, driven largely by the Construction Innovation Hub's activity in this area. As this book intends to show, there is little doubt that a shift towards a more manufacturing-led approach to the built environment can lead to a significant increase in the quality, efficiency and quantity of output from the construction industry. Product platforms provide one approach to achieving this if developed and applied effectively and with careful consideration of their suitability and scope. Mott MacDonald (2023) estimates that product platforms can reduce construction costs by up to 31% for certain asset types, worth up to £1.8 bn a year to the UK Government's social infrastructure spending alone. This can provide a multiplier effect to increase real UK GDP by up to £7.8 bn a year. At the same time, they can provide wider national social value and levelling up by enabling the redistribution of construction-related jobs to communities that would otherwise miss out due to their distance from construction demand. Such regional manufacturing hubs can provide safer and more stable employment relative to traditionally transient project working and help create centres of manufacturing excellence. So, why hasn't this approach been used in the past?

First of all, the level of investment in terms of time and money to create a product platform must not be underestimated. It is a significant undertaking and, in a construction industry that is traditionally project-based, the capability and willingness to invest in such platforms are not common without clear sight of a suitable, ready-made pipeline to address. A lot has changed since the postwar period of rebuilding using prefabricated units, which coincided with the modernist functional movement leading to mass production of largely identical homes. Today, customers across industries are more likely to demand products (or built assets) tailored to their requirements, and buildings are no exception. As such, buildings are traditionally (and rightly so) typically designed around the needs of their users (or at least user types), not around the capability of a particular construction method, leading to every building being largely unique. But that does not preclude manufacturing-led construction; as Thompson (2007) said, 'manufacturing-led construction should facilitate great architecture, not define it'. Platforms provide the opportunity to deliver unique assets without significantly limiting production efficiencies. In other words, a well-developed and considered platform capable of delivering mass customisation can largely create its own pipeline instead of relying on a ready-made pipeline in any given sector or from any client. This is key to the development of a sustainable platform, with examples including the entities platform highlighted in Box 13.2 or Platform II described in Box 13.3. Relying on a predicted pipeline of Government-led projects alone may not be sustainable as priorities change along the political cycle, and pipelines relying on a single channel or sector alone may be at risk of sudden political or market change that make them unstable sources of demand. This is a key point; for a platform to be effective over the long term it is likely to need a significant pipeline and a supportive supply chain. However, the same can be said for other approaches to delivery that may not have the same set-up costs.

Even with a pipeline in place, there needs to be the desire and capability to develop a platform in the first place, and who is likely to do that? In principle, any organisation can develop a product platform, but it does need to be an organisation with access to relevant and potentially significant resources. However, once a platform is developed, it can be accessible to businesses of any size as suppliers or even as users, depending on whether it is an open (open to any supplier) or closed platform (with a preapproved supply chain). In an ideal world, there would be one platform for construction that enables multiple suppliers to feed in to interchangeably without significant impact on cost, rework or production (think of the USB-C connection for devices compared with each platform provider having their own connector) but, realistically, the industry is too diverse for that to happen. It is more likely that many platforms, both open and closed, will develop and often compete, but there will also be a significant proportion of construction work that is still delivered using traditional construction methods or other forms of manufacturing-led delivery.

Box 13.2 Entities platform

The entities platform is an approach developed and used by the author to define and create product platforms which has been used in the development of several platforms over the last 15 years. The starting point is to assess the characteristics of different asset types included in the Entities table of the latest version of Uniclass or another recognised system. The characteristics typically considered include

- form
- function
- performance
- specification
- amenity
- brand.

These are defined in detail and enable the characteristics of assets to be understood across traditional sector boundaries. In addition to these characteristics, properties such as common spans, building height and any other relevant bespoke characteristics are assessed. This results in potential markets for a given technology being defined – for example, structures with long spans that have high amenity and brand value. Once defined, the specific markets are analysed in terms of how they are currently delivered and, using the Five Ps model, alternative systems and delivery methods are developed that enable delivery of multiple asset types efficiently and effectively while maximising value to the customer.

However effective a platform is, it will not usually be the case that a platform and its peripheral systems can deliver complete projects, with ground works and potentially bespoke elements requiring more traditional construction methods (for example, a podium in multistorey developments). However, it is likely that there will be knock-on benefits for the bespoke elements of projects using platforms, such as less movement on site, improved tolerances and interfaces with the platform elements and clearer system boundaries than on a traditional site.

Even with bespoke elements being carried out traditionally, the use of product platforms significantly increases the amount of project-agnostic work that is carried out in advance, such as platform design. This leads to significantly less instance-specific design, as illustrated in Figure 5.2, and also has an influence on that element of design. For example, using a manufacturing-led approach to structural frames is likely to lead to a much more accurate frame, meaning that the design and on-site application of façade systems can become significantly easier and more repetitive due to the lack of dimensional variation. In addition, it is unlikely that the platform mindset of improving efficiency and reducing waste and material use is not spread across a project. As explained in Chapter 5, a manufacturing-led mindset to construction delivery can be liberating and lead to efficiencies being found and explored beyond the initial area of focus.

In moving design to a project-agnostic model, the timing of interactions and decision making on solutions must be considered. For the application of a platform to a project instance to be effective and to minimise abortive work, it is important for the platform approach to be considered early in the project cycle; in fact, as early as possible. Without that early engagement to clearly define and understand drivers and applicability of a given platform, the chances of success are reduced. So, moving engagement of the delivery team to the earlier stages of a project means a different approach to procurement and potentially earlier decisions on detailed requirements and solutions. This earlier engagement of solution providers does not suit every client or every project, and is a strategic choice that must be clearly understood before a platform approach is utilised.

It is important to recognise that project-agnostic solution development does not mean that production is necessarily moved off site. The priority is to create the most efficient processes possible given a set of requirements, which may include the desire to make use of local labour and not just rely on production at remote factories, potentially breaking the relationship between production and the local community. Large-scale production off site may also lead to less small and medium-sized enterprise (SME) involvement unless they have access to the aggregated supply chain.

Box 13.3 Platform II, Bryden Wood

Platform II identified features such as floor heights and structural spans that are common across a range of assets, and a kit of parts solution was developed that is capable of delivering those assets – for example, schools, apartments and hospitals. The platform utilises standardised connections but has variable heights and spans, with a common range of 6–9 m, a range that Bryden Wood estimate can be used to deliver approximately 70% of public sector procurement (Bryden Wood, 2023). The kit of parts is used to deliver superstructure, façades, services and fit-out. The kit of parts uses predominantly readily available products, potentially used in new ways to enable high levels of productivity, irrespective of whether work is carried out on or off site. The process is what is most important, not the location.

Platform II was utilised on The Forge (completed in 2023), a commercial development in central London. The solution achieved a reduction in steel of over 18% compared to a traditional solution, a potential programme reduction for the superstructure of 25–55% and potential improvements for mechanical and electrical (M&E) services of 66–90%.

Generational platforms are not common in construction at present but, as is explained later in this book, an ongoing relationship between manufacturers and their products offers significant opportunities for all. So, while limiting a platform to the construction phase is a good start and can lead to significant improvements, it is limiting these opportunities, and so construction should not be seen as the final step in a platform journey.

13.3. The platform rulebook

Following on from the *Construction Playbook* (HMG, 2020) and its support for a platform approach, the Transforming Infrastructure Performance (TIP) roadmap was published in 2021 (IPA, 2021). This stated that the Government will, through a platform approach, 'generate societal outcomes from its pipeline, by enabling a disaggregated manufacturing industry that creates stable and inclusive employment where jobs are most needed'. This includes an intent 'to support a future mandate for construction platform approaches for relevant assets'. These reports signalled the Government's clear intentions to support a platform approach to deliver public projects, but a common understanding of what a platform was or how to develop one remained unclear. To resolve this, in 2022 the Construction Innovation Hub published the first version of its platform rulebook, which has since been updated (Construction Innovation Hub, 2023a), which intends to support those looking to develop a product platform and provides a series of guiding rules and principles. While product platforms can certainly exist outside of those rules and principles, the rulebook provides a thorough understanding of the concept and a good foundation for future platform development. It describes platforms as sharing common attributes

- a set of low variety, common assets shared by a number of products. As described earlier, these assets can be common products, processes, knowledge, people or even plant
- a complementary set of peripheral assets that can enable mass customisation
- a stable and common interface enabling the link between core and peripheral assets
- a set of rules or standards governing how assets can be integrated.

The rulebook is supported by a series of further documents called the Platform Papers (Construction Innovation Hub, 2023b) which provide further guidance on implementation, as well as by a quality approach known as construction product quality planning (CPQP) (Construction Innovation Hub, 2022). This tailors a number of tools and methodologies used in manufacturing to product development in construction.

The Seismic Platform® described in Box 13.1 and Platform II described in Box 13.3 both provide real-world examples of the application of the principles set out in the platform rulebook.

13.4. Summary

Product platforms have been around for some time in many manufacturing industries but are a relatively new concept in the built environment, especially for those not involved in product development. They provide a means to move away from continuously investing in instance-specific design development on a project-by-project basis towards more project-agnostic development of common assets that can be used multiple times, without creating cookie-cutter buildings across the landscape. In short, they can help deliver the vision of mass customisation – the delivery of customised built assets using efficient solutions that can be mass produced, whether that be on or off site. Platforms can provide opportunities for suppliers of products and services large and small to feed into supply chains, whether that is to provide standardised solutions or peripheral products and services.

A platform approach to delivery is supported by Government, and delivery of many public assets is likely to utilise this approach in the years to come. In creating repeatable processes and products, platforms can open the door to the effective adoption of automation, which may otherwise be harder to achieve on individual, bespoke design and delivery.

REFERENCES

Bryden Wood (2023) *Platforms in Practice*. Bryden Wood, London, UK.

Construction Innovation Hub (2022) *Construction Product Quality Planning Guide*. Construction Innovation Hub, London, UK.

Construction Innovation Hub (2023a) *Product Platform Rulebook*. Construction Innovation Hub, London, UK.

Construction Innovation Hub (2023b) *Platform Papers*. Construction Innovation Hub, London, UK.

HMG (Her Majesty's Government) (2020) *The Construction Playbook*. The Stationery Office, London, UK.

IPA (Infrastructure and Projects Authority) (2021) *Transforming Infrastructure Performance: Roadmap to 2030*. Cabinet Office, London, UK.

Mosca L, Jones K, Davies A, Whyte J and Glass J (2020) Platform thinking for construction. *Transforming Construction Network Plus, Digest Series*, No.2. Transforming Construction Network Plus, London, UK.

Mott MacDonald (2023) The Value of Platforms in Construction. https://www.mottmac.com/download/file?id=45219&isPreview=true (accessed 01/08/2024).

Thompson S (2007) Dwellings for today and tomorrow: A people-focussed, sustainable approach to design utilising an open building manufacturing approach. In *Open Building Manufacturing: Core Concepts and Industrial Requirements* (Samad A *et al.* (eds)). ManuBuild, Espoo, Finland, pp. 49-68.

Section 4

Systemise

emerald PUBLISHING   ice Publishing

Steve Thompson
ISBN 978-1-83608-599-7
https://doi.org/10.1108/978-1-83608-598-020241017

Chapter 14
Premanufacturing and integration models

14.1. Introduction
Once a project's desired outcomes have been clearly defined, work can begin on identifying the most appropriate delivery method. It is important to remember that the use of modern methods of construction should not be an end in itself but a way of achieving identified outcomes and, if a delivery method is chosen before the outcomes are clearly defined, then it is unlikely to succeed.

The purpose of this chapter is to explain the implications of premanufacturing and different models for integrating it into the delivery and operation of assets to maximise the value that can be delivered. In recent years there have been a number of high profile off-site business closures (along with many more traditional business failures), and these can often be explained by heavy investment in production facilities that rely on significant demand pipelines to remain profitable, which did not materialise. In some cases, businesses look to integrate different elements within the supply chain to make delivery more efficient and predictable, but still rely on a constant and often significant pipeline of projects to succeed which can be hindered by many of the usual factors influencing construction, such as change in demand and delays due to financing or planning. This chapter looks at how solutions can start small and be scaled to reduce reliance on significant upfront investment and continuous pipelines coming from a traditionally fragmented market. This is why markets that see the highest adoption of premanufacturing tend to be those with high demand and low availability of necessary skilled and affordable labour, such as the south east of England and London in particular.

The ability to start small and scale up or down as needs change is important for long-term success. However, to date, much of the guidance on using different forms of premanufacture tends to rely on consistently high demand to get economies of scale and repeatability, and also to justify the upfront involvement of manufacturing teams in a project timeline to ensure projects are designed to be compatible with premanufactured solutions. However, by using the Five Ps model described in Chapter 3, alternative integration models can be developed which enable scale across projects and sectors without over-relying on consistently high demand for a particular model of asset within a particular market. AUAR, described in Box 4.1, also focus on micro factories to minimise the upfront investment needed.

14.2. Premanufactured value
There are a number of ways that the level of premanufacturing on a project can be described or assessed, and the most commonly used is premanufactured value (PMV). This is simply a measure of the value of work carried out off site as a proportion of the total value of work on a project. More specifically, it can be calculated as a percentage by

$$\frac{\text{Off-site materials} \pm \text{labour} \times 100}{\text{Total materials} + \text{labour and prelims}}$$

The idea of PMV is that it is a relatively simple way of describing how efficient a project is by illustrating how much work is carried out off site. However, it needs to be used with caution, as building off site is not always the most efficient method of delivery. For example, if a premanufactured solution doubles in price relative to on-site activities due to a supply chain issue, using the PMV logic the project is suddenly twice as efficient, which is clearly not the case. Also, especially for projects with little repeatability, the cost of premanufactured solutions can be greater than traditional approaches (for example, Constructing Excellence (2021) estimated the cost of modern methods of construction (MMC) to be in the region of £3000 per m², whereas they expect it to fall to approximately £2000 per m² once the industry scales up production). The higher the PMV, the greater the amount of work that is carried out off-site, so in theory if all of a project was carried out off-site (which is not possible), then the PMV would be 100%. A further challenge with PMV comes in refurbishment projects; if an existing building system is reused on site, it is likely to be more efficient than manufacturing and installing a new system, but this will not be recognised as off-site and so will not contribute to the PMV percentage.

The Construction Leadership Council (CLC) set the benchmark for PMV at 40%, so anything above that is seen as an improvement on business as usual. However, to give a more rounded view of premanufacturing, the CLC also have other key performance indicators (KPIs) relating to modern methods of construction, which cover

- pre-manufactured value (PMV)
- productivity
- capital cost
- prelims cost per home built
- days on site
- quality rating
- waste generated
- embodied carbon
- energy performance certificate (EPC) rating (Construction Leadership Council, 2018).

These give a much more holistic view of the adoption of manufacturing-led construction.

While PMV does provide a high-level indication of efficiency in most circumstances, it needs to be understood that moving to a premanufactured model can initially lead to longer design phases as design teams familiarise themselves with designing for manufactured solutions. At the same time, work needs to be carried out earlier in the process to ensure that manufacturing can be based on fully resolved designs, as late changes are likely to be cost prohibitive or lead to abortive work.

Another example of a more rounded approach to assessing the potential efficiency of premanufactured solutions is the MMC toolkit developed by the NHS and its partners for their Procure 23 framework (ISG, 2023). The toolkit does have a target of 50% for PMV, but also has an overall target of 70% for MMC for new build and 50% for refurbishments. To calculate the overall MMC percentage, scores include a percentage for category 7 solutions (on-site efficiencies) and a new category 0 which covers the briefing, scoping and design phases of a project, including levels of standardisation. Category 0 is an enabler for successful implementation of premanufactured solutions and covers areas such as standardised floor heights, building grids and components, so provides a good indication of potential efficiencies being made.

14.3. Production methods

While considering the broader standardisation and rationalisation of designs along with on-site processes gives a clearer picture of the level of manufacturing-led construction than just PMV alone, the method of production is also required to give the full picture. The use of PMV, category 0 and category 7 do not necessarily cover how a building system is produced or how replicable it is across projects. For example, simply moving construction to an off-site location and carrying out construction in the same way is unlikely to have a significant impact on the outcome of a project, whereas if systems are assembled on a production line in a manufacturing environment the impact can be significant. To address this, the Manufacturing Technology Centre (MTC) developed the Capabilities for Modern Construction (CMC) framework (MTC, 2022). The framework has seven levels, each with an increasing level of production and scale.

- Traditional construction
  - o Level 1: traditional construction process – on-site build that does not use prefabricated componentry
  - o Level 2: advanced construction process – on-site build that uses some prefabricated components
- Modern methods of construction
  - o Level 3: off-site static modular construction process – static off-site batch manufacturing where the units remain still and workers move around the units (this is the level that most UK off-site businesses are at)
  - o Level 4: conversion of the off-site construction process into an off-site manufacturing process – a line manufacturing process for up to 1500 units per year, with a high level of process standardisation
  - o Level 5: low volume manufacturing process for off-site construction – for up to 3000 units per year, assembly is mostly based on pre-assembled components
  - o Level 6: medium volume manufacturing process for off-site construction – for up to 6000 units per year and based on a continuous flow line assembling pre-assembled components
  - o Level 7: high volume manufacturing process of off-site construction – for approximately 10 000 units per year based on a continuous flow process and assembly of pre-assembled components.

Figure 14.1 provides an indication of the impact of different production methods for each level of the MMC framework in terms of reducing the need for, or eliminating completely, tasks that would

Figure 14.1 Impact of production on levels of automation (Author's own)

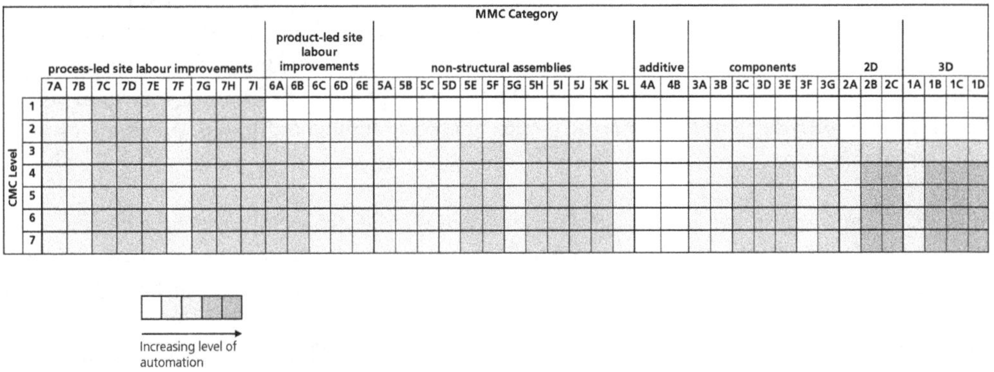

otherwise need to be carried out using traditional methods. The figure is based on the percentage of tasks that can be automated, not a percentage of time saved. The figure results from primary research carried out for this book and is described in the Appendix. As is to be expected, the higher the level of production capability the higher the level of automation. Likewise, the greater the level of prefabrication, the higher the level of automation, but that does not necessarily mean that volumetric leads to more automation than panelised systems.

The CMC framework is based on a matrix of core business capabilities and attainment levels and can be used to develop production strategies for the use of manufacturing-led solutions. The MTC believe that operational efficiency can double for each level a production moves up the scale, so the potential for manufacturing-led construction is clearly not about just moving construction off site; it is about doing things in a more structured and repeatable way. It is important to note that there is a balance that needs to be made between the level of production and investment made with the available demand to support such capabilities. The following sections will highlight examples of different approaches being taken to strike this tricky balance.

14.4. Types of integration

When looking at the traditional construction supply chain, it is clear that it is incredibly fragmented. Even where precision-engineered products are used, they often pass through several layers of the supply chain before being installed in the final asset, from manufacturers to distributers, and from subcontractors to main contractors. At the same time, often many products and systems come together in a single construction or assembly phase from a number of different sources. When looking at manufacturing-led or industrialised delivery, there is likely to be integration of these processes to minimise interfaces between products and actors and to reduce risk and on-costs. There are different kinds of integration, as highlighted in Figure 14.2, which can be summarised as

- vertical integration
- horizontal integration
- multidirectional integration
- life cycle integration.

The manufacturing outputs that are typically used in production strategy are delivery (lead time and reliability), cost (product cost per delivered unit), quality (conformance to product specification), performance (features and product performance relative to similar products), flexibility, (volume and product mix) and innovativeness (ability to introduce new or modified products). All of these hopefully act together to produce the desired project outcomes, and they will also have an impact on the appropriate level of premanufacturing and production capabilities and repeatability of solutions. Not all projects are suitable to feed large-scale production facilities, and so to provide industrialisation across smaller or more complex projects requires a different approach to large-scale, repeatable assets.

Another recent trend is for an approach which overlays the different types of integration and is driven by intellectual property and added services. In this form of integration, which is an asset-light approach, the integrator takes the form of a common platform provided by the platform owner which can be used in vertical, horizontal or multidirectional integrations. Examples of this include the Seismic Platform® (described in Chapter 13), AUAR (described

Figure 14.2 Integration models (Author's own)

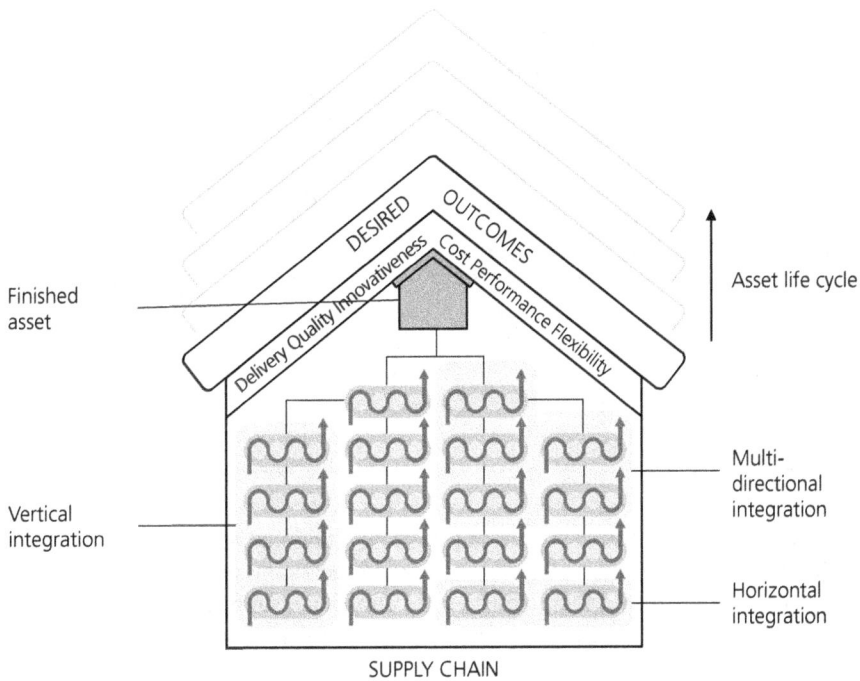

in Chapter 4) and Modulous (described in Box 14.2). One of the most significant benefits of this approach is that there is no reliance on a significant fixed factory asset that needs to be fed. Instead, there is flexibility of supply chain partners and the ability to scale or reduce in size at any point, and even the ability to change supply chain partners as circumstances dictate. This level of flexibility also enables the location of facilities to vary depending on demand, which has the added benefit of potentially reducing transportation costs and having regional hubs that support local economies.

14.4.1 Vertical integration

Vertical integration is perhaps the form of integration that most often springs to mind when discussing integration of supply chains, but it is by no means the only kind. Vertical integration is where a business extends their offering to include the next or previous role within their supply chain. For example, a supplier of light steel studs may move to offer panelised or even volumetric systems using their sections, or provide closed (boarded) panels instead of studs alone.

Vertical integration can happen in either direction. For example, as illustrated in Figure 14.3, a main contractor may move to provide their own modular systems as Berkeley Homes and Laing O'Rourke do. Alternatively, manufacturers may move to provide construction services to support their system offering, such as Genuit Group and their prefabricated pipework systems. Equally, new businesses can pull together products and services that would traditionally be delivered by separate businesses, such as TopHat.

Figure 14.3 Vertical integration (Author's own)

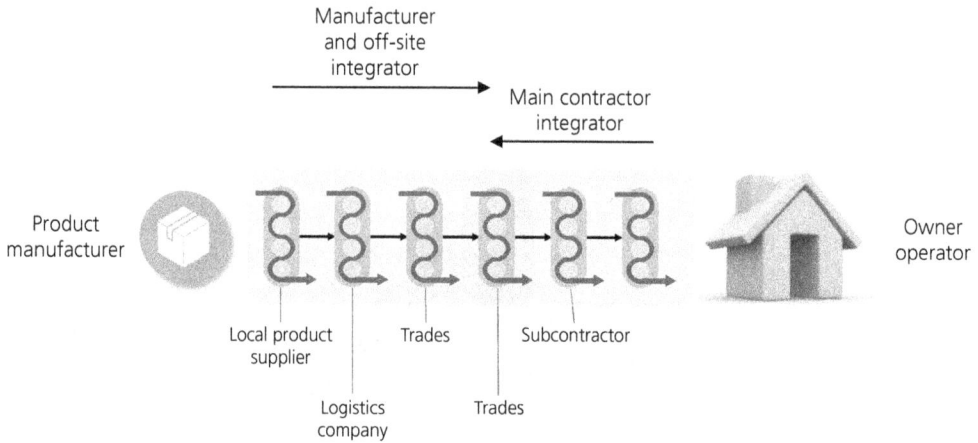

Vertical integration may also take the form of providing additional services relating to a product, which would otherwise need to be offered by others in the supply chain. An example of this is where windows are installed by a supplier instead of only being supplied for others to install. This is known as servitisation, and an extreme version may be providing a home with guaranteed zero energy bills instead of providing an energy efficient building alone, as illustrated in Figure 14.4. Productisation is where services are packaged into products, such as design services provided as a lump sum for a clearly defined package of work and doing this on a repeatable basis instead of simply charging by the hour.

Figure 14.4 Servitisation and productisation (Author's own)

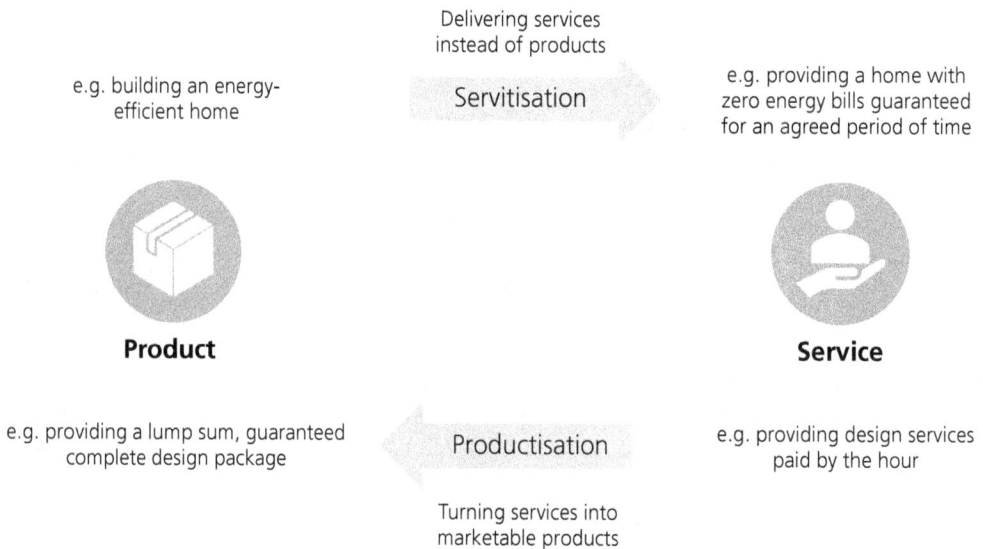

The use of subassemblies for items such as door sets are increasingly common in construction, as is the use of panelised building systems. Still, according to a survey by NHBC (2016), in 2015 only approximately 15% of new housing was built using any form of off-site construction, so the industry still has a long way to go.

So, why do businesses look to vertically integrate? One simple reason is to grow a business by providing additional services that would otherwise be provided by others, which has the added benefit of having a more direct relationship with businesses further up the value chain and thus having more influence on decisions that affect the business. It can also create a larger, more consistent pipeline for a business's core business. Operating in the reverse direction, a contractor can gain more certainty and confidence in their supply chain by integrating key packages of work that their business relies on.

14.4.2 Horizontal integration

Horizontal integration is far less common than vertical integration and involves the coming together of products or services that would otherwise be parallel in the supply chain. A simple example is a contractor providing and installing a package of doors and windows instead of just doors, or where a business provides both walls and floors instead of only one or the other. A more complex version is PT Blink, described in Box 14.1.

Horizontal integration can be harder to achieve than vertical integration, as it relies on either using existing capabilities to extend the reach to related packages of work or obtaining new capabilities. The value of horizontal integration is that it can de-risk a business's reliance on other suppliers of products and services that may otherwise have an influence on core business – for example, where a package of work is only let when all suppliers within that package are ready to deliver.

14.4.3 Multidirectional integration

Multidirectional integration is where integration occurs both horizontally and vertically. A typical volumetric business delivering finished modules is an example of this form of integration. Modulous is an asset-light example; a number of subsystems were assembled off site in local hubs into finished volumetric units ready for delivery and installation.

To provide additional flexibility to cope with different project or site constraints, some providers enable delivery in either 2D panels or 3D modules without significant changes to the end product, depending on the level of prefinish that is required. An example of this concept was developed in the ManuBuild research and development project, a 6th Framework project funded by industry and the European Commission between 2006 and 2009. A system was developed that enabled construction as 1D components, 2D panels or 3D modules, and that flexibility enabled different building types and configurations to be delivered using the same components, thus significantly widening the market for such a solution.

It is important at this point to emphasise again that integration does not always mean integration of physical products into systems. Using the Five Ps model, it can be any of those, or all of them that become integrated. An excellent example of systemised integration can be found in Mace's High Rise Solutions (HRS), where success relies on the coordination and careful design and implementation of processes and products to deliver the final results. Mace report up to 25% faster delivery, 40% less transport and 75% less waste than traditional high-rise construction methods as a result (Mace, 2023).

Box 14.1   PT Blink

PT Blink's design, manufacture, integrate (Blink DMI®) methodology is based on a patented flat-packed posttensioned structural steel frame which is manufactured off-site by accredited partners and then installed on-site using accredited integrators. Because the system is dimensionally accurate and designed for manufacture and integration, it is significantly quicker to install (around 50%) than traditional methods for high-rise construction and means that the façade and building services systems can also be installed as the frame goes up. To enable this to happen, the building services and façade systems are delivered in parallel to the structural frame, meaning that the PT Blink network of suppliers is effectively horizontally integrated. Figure 14.5 illustrates a project using PT Blink.

Figure 14.5   PT Blink's patented flat-packed posttensioned structural steel frame (Source: PT Blink)

Box 14.2 Modulous

Modulous took an asset-light and kit of parts approach to their offer. They designed a kit of parts that rationalised over 1300 standard components into less than ten subassemblies, which were then assembled into volumetric units. Modulous did not own fixed production assets; instead they had a network of approved suppliers who supplied and assembled different elements along the supply chain. The final modules were then assembled as close to the final site as possible, meaning that transportation was optimised to minimise transporting large 3D modules over long distances, but also supported local employment. The final modules were assembled onsite into one-, two- or three-bedroom homes which were configured based on their proprietary TESSA software. The asset-light approach meant that the solution was easily scalable, but also that new supply partners could come on board as the offer and market continued to develop, while not having a fixed capital asset reduced the reliance on a constant and significant pipeline to continually keep the production line running. Figure 14.6 illustrates the Modulous development process.

Figure 14.6 Modulous development process (Source: Modulous)

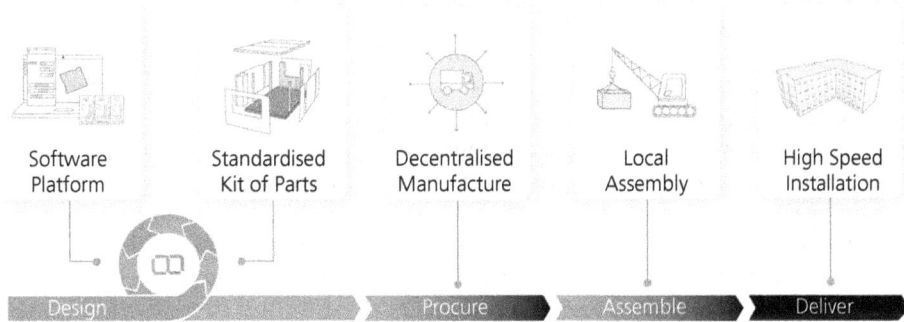

| Software Platform | Standardised Kit of Parts | Decentralised Manufacture | Local Assembly | High Speed Installation |

Design — Procure — Assemble — Deliver

14.4.4 Life cycle integration

Life cycle integration is the least common form of integration in the built environment today, but with the significant shift in recent years to whole life value, it is likely to become more widespread. However, in some areas of the industry it is commonplace; in shopping centres and commercial offices, for example, where the main building and services are provided by the landlord, but the fit-out of individual units is carried out by the business leasing the unit. Once that business is replaced with a new one, the building is designed in such a way that it can be readily converted to a new configuration; it is designed with future uses in mind, not as a one-off fixed unit with only one finite use. This can be described as in-use adaptability, but in this section the focus is on situations where a business maintains a relationship with its product or system throughout the asset's life cycle beyond traditional product warranties. This form of life cycle integration is less common than the in-use adaptability model, but forms part of an industrialised built environment, and can be a whole building such as the Adapteo, or can be a servitised solution where the client buys performance instead of a product (for example, guaranteed energy performance instead of a product alone).

Life cycle integration may seem alien to many in the construction industry, who are typically focused on delivering new assets without significant consideration of future use or reconfiguration. However, the open building movement has been around since the middle of the twentieth century and focuses on the ability of buildings to be readily reconfigured or upgraded with minimal effort. With the application of automation, digitalisation and in particular industrialisation, this vision is inevitably more achievable today, if considered from the outset. Creating such adaptable assets is not so much about designing in flexibility as it is about designing out inflexibility. Once the future adaptability of an asset is considered from the outset, a new mindset can emerge that is much more akin to product design than it is to construction detailing, and the timing and order of construction activities becomes increasingly important.

There are two main areas to cover in life cycle integration. One is creating new assets that are readily adaptable in the future. The other area looks at how we can use manufacturing-led solutions to adapt existing assets today and again ensure they can be adapted further in the future.

The main area where industrialisation is likely to take place is undoubtedly in the delivery of new assets; starting with a relatively blank canvas is always going to be easier than interfacing with the unique configuration and measurements of existing assets. However, with the rapid development of digital engineering, and in particular of measuring and mapping existing assets, the interface between precision-engineered products and unique existing assets can be easier to manage. For example, with laser scanning achieving accuracies of less than a millimetre, the production and installation of manufactured systems no longer need rely purely on on-site cutting or shaping of systems to fit. Panels can be predesigned and shaped to fit perfectly, and often undulations in surfaces can be dealt with through the design of adaptable subsystems. This opens up a significant market opportunity for businesses that would otherwise not consider the refurbishment sector as a possibility. By creating a potentially direct relationship between the system provider and the end user (the occupant), there is potential for a longer-term relationship, even to the extreme of product leasing instead of direct sale. This direct relationship relies on the separation of design responsibilities – in other words, providing the infrastructure of a building and allowing the occupant or user to fit out or configure their own spaces as they desire.

In the Netherlands, Energiesprong is an excellent example of the application of industrial prefabrication to the renovation sector, and is now growing in the UK, as described in Box 14.3.

In addition to the industrialised refurbishment of existing buildings, it is crucial to explore how new buildings are designed and delivered. As discussed in Chapter 3, if buildings continue to be delivered in the same way they have been in the past, future generations will surely need to significantly refurbish, remodel or replace these relatively new buildings before they reach their design life in order for them to perform and support the lives of their users for years to come. An industrialised built environment's strength potentially lies in the direct contact between a system provider and its user, in a similar way to the relationship between a car manufacturer and its driver. In this example, the car can be customised based on individual demand using an integrated system of systems. The difference in an industrialised built environment, however, is that elements of the system will be replaced and reconfigured over time.

The coordination of so many products into a larger composite system is based on the simple principle that standardisation must only deal with interface conditions. Where products of two producers

Box 14.3 Energiesprong

Energiesprong, originally developed in the Netherlands, is a significant industrial prefabrication model for whole-house net-zero renovation projects, with over 5700 dwellings already renovated to date. It includes an integrated prefabricated façade and roof system in conjunction with preassembled 'energy pods' to renovate predominantly social housing units to create net-zero or even energy-positive dwellings. There has even been legislation put in place to allow landlords to charge an 'energieprestatievergoeding' (energy performance renumeration), which a tenant pays the landlord in return for guaranteed energy savings. Energiesprong UK is now working to scale up the model in the UK, and clearly demonstrates the potential for manufacturing-led solutions in refurbishment. Figure 14.7 presents a successful project in Nottingham.

Figure 14.7 Energiesprong project designed by Studio Partington and delivered by Melius Homes for Nottingham City Homes, Nottingham (Source: Energiesprong UK)

meet, conventions of details and dimensions must be established. Beyond that, the design of individual components or systems can vary as long as they meet the overall requirements of a system. An example of such an approach is the domestic kitchen. There is the design of the system (standardised unit dimensions and interfaces), and then there is the unique instance of that system, which is designed and configured to the space and user requirements. An industrialised built environment, or lifestyle integrations, will no doubt develop within the next decade as the constant march towards whole life value and technology develops. Box 14.4 provides an example of a building designed with future adaptability in mind from the start.

Box 14.4 NEXT 21, Osaka, Japan

NEXT 21 is an experimental project using systemised construction to provide a core and fit-out model for housing. The apartments are designed by 13 different architects within a system of coordinating rules, with service distribution spaces in both the floor and ceiling to maximise flexibility of layout. The building façade was designed as an independent system, providing flexibility in how individual units are finished. One of the units has since been extensively remodelled, and all work was carried out from within the unit, minimising disruption for others. It is worth noting that, in the UK, planning is likely to be an issue when it comes to providing such flexibility for individual units, and this will be discussed further in Chapter 17.

14.5. Summary

Premanufactured solutions can vary significantly in terms of the level of work carried out in a factory against on site, but also in terms of the level of production capability used. The majority of the off-site industry in the UK still operates based on the workforce moving around modular units, not the units being manufactured moving between stations, and there are benefits to that approach; it can mean there is flexibility to operate on projects with smaller volumes of repetition, for example. However, if the construction industry is to take full advantage of the benefits that can be delivered through prefabrication and manufacturing-led solutions, there needs to be a shift towards more automated production capabilities while still providing a level of customisation. In short, a move towards mass customisation. To achieve mass customisation, and to minimise the risk of a supply chain failure, standardisation can play a significant role. The question is, what level of standardisation is most effective? Is it at a building or space level? The length or height of modules? The answer to provide the most flexibility is likely to be standardisation of interfaces in the first instance, and potentially standardisation of key building dimensions, such as grids or floor to ceiling heights. This will increase repeatability and confidence while maximising design flexibility, and will be discussed in more detail in Chapter 17.

In recent years there have been a number of high-profile business closures in the off-site sector (as well as in the wider construction industry). There are a number of reasons why any business can fail, many relating to issues outside of the core business model or technology adopted. There have been just as many success stories over the same period, and so it is important to carefully consider what technologies are most appropriate, the speed at which a business can scale sustainably, and what core business model (or models) are most likely to succeed. It is important to not jump to a solution before first going through the definition of need and to undertake thorough market analysis before developing any new business in such a rapidly changing construction industry.

REFERENCES

Constructing Excellence (2021) *Benefits of Modern Methods of Construction in Housing: Performance Data and Case Studies*. Construction Excellence, Watford, UK.

Construction Leadership Council (2018) *Innovation In Buildings Workstream Housing Industry Metrics*. Construction Leadership Council, London, UK.

ISG (2023) New NHS toolkit for MMC to drive cultural change across UK healthcare. https://www.isgltd.com/uk/en/news/new-nhs-toolkit-for-mmc-to-drive-cultural-change-across-uk-healthcare (accessed 01/08/2024).

Mace (2023) *Mace Tech High Rise Solutions.* Mace, London, UK.

Manufacturing Technology Centre (2022) *Development of Capabilities for Modern Construction (CMC).* Manufacturing Technology Centre, Coventry, UK.

NHBC (National House-Building Council) (2016) *Modern Methods of Construction: Views from the Industry.* NHBC Foundation. Milton Keynes, UK.

Section 5

Automate

emerald PUBLISHING · ice

Steve Thompson
ISBN 978-1-83608-599-7
https://doi.org/10.1108/978-1-83608-598-020241018

Chapter 15
Technology timeline

15.1. Introduction

As described in Chapter 1, before automating something it needs to be clear that the right thing is being automated, and that the requirements are clearly defined first. However, once all that has been done, there are a range of technologies that are already available to some degree that can be applied to automate tasks of many different types, which will be increasingly accessible and capable in the next decade. This chapter describes a number of these technologies and their potential applications in the built environment, and suggests a potential timeline for when they are likely to become widely adopted. The list of technologies is not exhaustive but is focused on technologies that will enable automation in some form. In Chapters 22 and 23, the application of these technologies, along with others, will form the basis of the analysis of the impact of automation on the future of work and on scenarios for the construction industry in 2035.

15.2. 4D modelling

This is the process of linking 3D models with construction, resources and programme information to virtually construct a project process by process to ensure it is feasible and optimal before real construction starts. It is described in more detail in Chapter 19. In the UK, Freeform are leading proponents of 4D modelling on complex projects. The technology is widely used today for projects of a certain size, but it is not expected to be easily accessible or used on small projects until later in the 2020s, if at all.

15.3. Additive manufacturing

Additive manufacturing, otherwise known as 3D printing, is the process of producing layer upon layer of material to create a 3D object. It is generally used on objects much smaller than buildings, and can be used to produce complex geometries (for example, from generative design) that would otherwise be difficult to produce. It has been used for many years to produce small prototypes of objects, but there are now examples where it is being used to produce single-storey building structures. Due to its limited multimaterial capabilities and size restrictions, it is always likely to be more useful at a component than a building level.

15.4. Advanced ground scans

Advanced ground scans use either ground penetrating radar (GPR) or electromagnetic waves to understand ground conditions and locations of buried pipes and structures. Leica produce scanners that can penetrate earth, concrete or other materials to find utilities across a site, limiting the need for disruptive surveys or damage. Emerald Geomodelling provide a 3D model of ground conditions and structures using a geo scan by helicopter, which penetrates up to 100 metres below the surface.

15.5. Asset management software

Asset management software is used to help asset managers better manage their assets, including both preventative and reactive maintenance. It can also help in managing asset use and be linked to sensors as part of a digital twin. It is described in more detail in Chapter 20. IBM's Maximo is a common asset management platform which uses artificial intelligence and sensors to optimise performance. On smaller, domestic projects it is likely to become more widely used in the short term, as smart devices for security, heating and power become more accessible.

15.6. Automated design

Described in Chapter 17, automated design automates repetitive design tasks, such as adding floors to buildings based on input parameters, or applies detailed products and systems to early-stage design models to create more detailed solutions. KOPE is an example of automation of the detailed application of products, and Autodesk Forma is an example of automated analysis of designs.

15.7. Blockchain

Blockchain is a distributed ledger technology that enables transactions or agreements to be stored across a network of computers, which cannot be changed once published. Potential uses include smart contracts, payment, procurement and supply chain management.

15.8. Building automation systems (BAS)

BAS are systems that can control asset-wide systems such as lighting, heating, ventilation and air conditioning (HVAC) and security access. They use sensors and controls to monitor and adjust systems. Linking BAS to machine learning can enable buildings to become programmable and responsive to user needs. BAS are described in more detail in Chapter 20.

15.9. Computer vision

Computer vision is a form of artificial intelligence where images are analysed and appropriate output created. Sort It AI uses computer vision to extract quantities from images of plans, and is described in Chapter 18. Computer vision can also be used to assess progress of construction projects through photographs, such as with Buildots.

15.10. Connected autonomous plant (CAP)

CAP is plant that is connected to its environment through sensors, and can move around a site automatically without requiring a human driver. In addition to movement, connectivity can enable locations of plant to be quickly understood, along with their relative performance. Built Robotics have developed a system for autonomous pile driving, and MachineMax have a platform for accessing and managing data from site equipment. The technology is expected to become common practice by the end of the 2020s, and is discussed in Chapter 19.

15.11. Connectivity platforms

Chapter 9 is dedicated to connectivity, and connectivity platforms act as a link between different disciplines and stages across the life cycle of an intervention, potentially from initiation through to handover. They enable data in different formats to be shared, federated and accessed by different stakeholders.

15.12. Design configurators

Described in Chapter 17, design configurators are used to create design solutions from a given set of products or systems. They are typically restricted to the capabilities of predefined systems, but can be combined with generative technologies to deliver more complex solutions with a greater range of technologies. Laing O'Rourke have developed a bridge configurator for their modular bridge system which configures single-span integral bridges based on geometrical and load input parameters.

15.13. Digital handover software

The traditional handover of a project from a construction team to the client or facilities management team can involve hundreds of documents and files in different formats, hard copy and digital. With digital handover software such as Asset Twin by Invicara, information is stored in an information model that can readily be accessed, shared and analysed in a digital facility manual. Models can easily be queried to find relevant components and associated information accessed.

15.14. Digital twins

Digital twins are not the same as 3D models; instead, they are virtual copies of a physical asset or process, with real-time connections to their physical twin. The principles are described in Chapter 5, and their application in Chapter 20, and they have several use cases including optimising energy use or procurement processes.

15.15. Drone scans

Aerial scans carried out by drones (otherwise known as unmanned aerial vehicles (UAVs)) can provide quick, unintrusive and reasonably accurate digital representations of sites before, during or after construction and can also be used to carry out visual condition surveys in otherwise inaccessible locations. Described in more detail in Chapter 19, data can readily be stored and shared digitally with reduced risk of human error.

15.16. Generative detailed design

Generating fully resolved detailed designs of assets is likely to be incredibly complex, but can mean that initial design solutions can be created and immediately detailed to provide confidence that a solution is deliverable. ArchiConDes have developed the RST technology with such capability, which is described in more detail in Chapter 17. Such technology can significantly reduce design and development time by automating design development. Once the technology is used in practice, it is expected to become widely adopted over a short period of time, and to some extent overtake generative form-finding design, as it is capable of the same functions and more.

15.17. Generative form-finding design

Most generative design tools used in the construction industry are restricted to early-stage optioneering and solution development, and generate solutions based on user input of parameters and site constraints. They are described in more detail in Chapter 17. TestFit provide such a solution to generate building configurations for apartment blocks and other building types, rapidly developing optimum layouts based on user input.

15.18. IoT sensors

Sensors are a key enabler of digital twins in that they can identify specific locations of objects, such as equipment or materials, the condition of an environment (for example, in concrete curing or internal space temperatures) and performance of systems (such as heating or air conditioning). Their use in asset performance is described in Chapter 20.

15.19. Laser scans

Laser scans can be used to survey the inside and outside of existing assets and can provide quick and accurate results which can readily be shared to mobile devices or computers. They are generally significantly quicker than manual surveys, but they can only survey what they see, so often need to be carried out from a series of different angles. The added benefit is that laser scans automatically create a digital representation.

15.20. Leak detection systems

These are used to find, analyse and quantify the size of leaks in water infrastructure. There are several ways that this can be done, but FIDO AI described in Box 20.2 is a great example.

15.21. Location systems

The purpose of location systems in construction is to accurately identify locations of objects or people on a construction site, not just in terms of horizontal positioning but also height above ground or floor level. For example, ZeroKey provide a real-time location system (RTLS) which monitors, visualises, analyses and reports locations in real time and so can be used for personnel and plant monitoring and measurement.

15.22. Logistics and supply chain software

Logistics software is used across multiple industries; however, as has been described in other chapters, the construction supply is typically more complex and fragmented than in other sectors. In construction, software such as WSCI (Wincanton Supply Chain Integrator) can be used to automate the visibility, tracking and analysis of complex supply chains, tracking each component. It can be used to help predict bottle necks, storage issues and forecast payment requirements.

15.23. Low-code platforms

Low-code is technology that allows people with little to no advanced software programming skills to develop applications that can be used to automate tasks. This opens up the application of automation to a significantly larger audience and means that users involved in any aspect of the built environment can automate repetitive tasks with basic skills using tools such as Microsoft Power Apps. It means that tasks can be automated where there may not be the market available to justify the expense of developing an app specifically for that purpose. Morta is an example of a low- or no-code platform specifically developed for the built environment and can be used to integrate different software packages.

15.24. Machine learning (ML)

ML is a form of artificial intelligence where algorithms are used to process large volumes of data and learn from it. Once taught, ML models then make predictions or identify patterns without being programmed to do so. In construction, ML can be used to optimise schedules, to predict issues and risks or predict asset performance. It can also be used for structural health monitoring, traffic flow prediction and many, many more applications, likely including many that have yet to be thought of. Optimise-AI use ML to optimise asset performance and energy use based on inputs from sensors and simulations.

15.25. Natural language processing (NLP)

NLP is a form of artificial intelligence where machines recognise text or voice, extract data and use that to produce an output. NLP can be used to extract requirements from documents or for inputting requirements into applications by way of voice input, thus quickly providing a digital requirement or instruction. It can also be used for relatively simple tasks, such as producing minutes of meetings or to power chatbots used on supplier websites to answer common questions.

15.26. Physical robots

Physical robots are more likely to be found on a production line than on a construction site, and that is likely to be the case for some time. However, site-based applications are beginning to grow, and Chapter 19 describes site applications such as surveying robots, painting robots and exoskeletons.

15.27. Predictive analytics

Predictive analytics involves extracting information from historical data and using this to predict future trends. nPlan use predictive analytics to learn from data on a significant number of previous projects to identify the risk of delay and to suggest potential improvements. Predictive analytics can also be used to identify future maintenance requirements.

15.28. Procurement software

Procurement platforms in construction can be used to publish requirements, obtain estimates and offers, and automate change management and payment through the supply chain. They are described in more detail in Chapter 18.

15.29. Programme optimisation

Programme optimisation includes 4D modelling for digital rehearsals of construction activities, machine learning for construction schedule optimisation (such as nPlan) and generative construction, where programmes are optimised based on a set of inputs and assumptions (such as ALICE Technologies, which links construction processes to BIM models). Synchro is another useful tool linking models with programmes. Solutions are described in more detail in Chapter 19.

15.30. Robotic process automation (RPA)

RPA is where preprogrammed software automates labour-intensive, repetitive computer-based tasks. It can be used to automate data entry, extracting data from numerous sources, to produce documents, such as invoices or reports, or to manage inventory levels.

15.31. Route optimisation

Route optimisation is the automation of designs and optioneering for linear assets, such as pipelines, highways or rail. It can be complex as it needs to not only consider design rules that impact on safety and performance but also cost, land use and environmental impacts. Optioneer by Continuum Industries enables users to incorporate geospatial information and generate multiple design solutions for infrastructure projects for review.

15.32. Safety software

There are many safety-related software solutions in the construction industry, the majority of which automate the collation and sharing of safety records and performance data. They can also be used to provide training. Some software uses computer vision to help identify potential risks.

15.33. Site communication

Site communication is clearly key to the successful, safe and effective delivery of construction projects and, as such, there are a number of software solutions focusing on this area. JobNimbus is an example of this type of software, which enables users to assign tasks and work orders to specific teams or individuals and to track the status of those tasks along with providing a clear communication route. Solutions are discussed in Chapter 19.

15.34. Site management and reporting software

These tools help provide analytics on performance, but also automate the reporting, collation and presentation of progress and other key data. Sablono, for example, enables individual jobs to be identified, assigned, tracked and reported as being complete, and shares this information with relevant stakeholders across the platform. Solutions are discussed in Chapter 19.

15.35. Smart contracts

Smart contracts are contracts that are human- and machine-readable. They can support automation of contract management – for example, by triggering payment once notification is received that an output has been delivered. Contracts can be automatically updated to incorporate agreed changes, and blockchain can be used to record transactions.

15.36. Supply chain digital twins (SCDTs)

SCDTs are models of the supply chain and all products and services within it, including location and state (for example, ordered, delivered, installed). As with all digital twins, they have real-time connections to their physical twin.

15.37. Waste management software

The management of construction waste is a significant issue for clients and contractors for projects of a reasonable size, and software can be used to make the process easier. Tools such as Qflow exist to automate the tracking of materials and waste entering and leaving a site through scanning of waste transfer or delivery notes and rapidly compare the results against project targets. The UK Government are planning to mandate digital waste tracking in the near future.

15.38. Workforce management software

Workforce management is clearly a key part of any construction project, not only for health and safety reasons but also to assess and manage performance and resources. Rhumbix is an example of a solution that tracks utilisation data for both people and plant, as well as change management.

15.39. Technology timeline

Figure 15.1 suggests potential timelines for the adoption of the different automation technologies described in this chapter and indicates level of adoption from initial applications to the technologies becoming ubiquitous. The level of adoption is influenced by much more than the availability, accessibility and capability of the technology itself; it is also impacted by common working practices in different industry sectors and relevance. For example, route optimisation is not relevant to vertical assets such as buildings and, as such, will never become ubiquitous, but it can certainly become common practice across a number of horizontal asset types. In a different way, site communication technologies are unlikely to be used at scale on small, domestic projects where the workforce is small enough to manage directly and the preference of traditional contractors is likely to influence the uptake of such technologies.

The forecast adoption rates and timescales are based on extensive desktop research, followed by discussions with those developing or implementing technologies today.

Figure 15.1 Timeline for technology adoption in the built environment (Author's own)

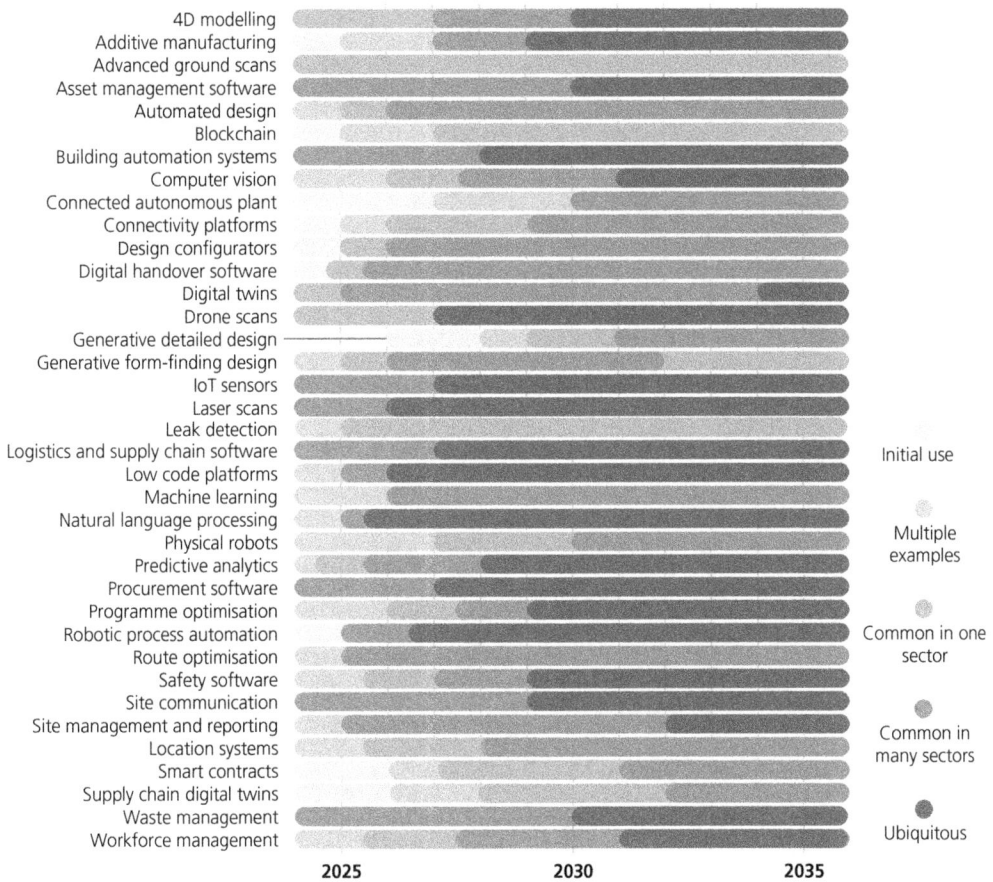

Some of the technologies described in this chapter are covered in more detail in other chapters, and all of them are considered as part of the implications on the future of work and future scenarios in Chapters 22 and 23, respectively.

emerald PUBLISHING ice Publishing

Steve Thompson
ISBN 978-1-83608-599-7
https://doi.org/10.1108/978-1-83608-598-020241019

Chapter 16
Business casing construction delivery

16.1. Introduction

Previous chapters of this book have covered the definition of need and requirements for interventions; these requirements then feed into an assessment of some form to determine what it will take to meet the requirements, and whether or not it is likely to be worth the time, cost and effort.

A business case provides justification for undertaking a project, programme or portfolio. It evaluates the benefit, cost and risk of alternative options and provides a rationale for the preferred solution (APM, 2019)

The construction industry has typically operated in a similar way to assess the business case for a development for centuries; a design team develops a concept that it believes meets the requirements, a cost consultant estimates how much it will cost to deliver and potentially a contractor or another consultant suggests how long it will take to realise. Based on that, the client considers whether the benefits outweigh the outlay. However, in the ever-changing world of today, this needs to be unpicked somewhat to ensure that the industry can deliver better outcomes and enable new technologies and methodologies to realise their potential to truly meet defined need.

Starting with defining the benefit, tools such as the Value Toolkit Construction Innovation Hub (2020) and methodologies such as the Green Book's Five Case Model (HMT, 2022) provide the means to clearly identify what the desired benefit is, while avoiding the need to design out potentially value-adding solutions without suitable consideration. When considering new construction technologies and methodologies, a clear, unrestricted definition of need is key to comparing solutions against more traditional approaches without prejudging how that need will be met.

Historically, projects have often been awarded based on a lowest cost basis, whereas the move to a more holistic approach of best value does provide more potential for new, alternative delivery models. The same flexibility is required when considering how and when procurement packages are defined. Many manufacturing-led approaches to delivery cross traditional package boundaries (for example, structural units combined with cladding solutions) and, as such, need to be considered accordingly. As outlined in Chapter 4 and Figure 4.2, manufacturing-led or industrialised construction changes the roles of and interfaces between different parties. This can have a significant impact on the management of risk but also on incentivisation of different players. Those delivering the majority of the construction activity (in Tier 2 and below) play a more visible role and may well play a more significant part in the development and assessment of the business case for an intervention. This requires as much of a cultural shift as it does a technological one, and means that new providers need to consider their own business case; are they looking at new delivery models that increase or decrease their risk? Are they entering new markets with new customer types, and do they have the breadth and mix of business models to enable them to repeat solutions across their target markets?

Finally, whether looking at the business case for a new delivery model for a business or for a project, it is important to make sure that the business case is live and considered throughout a project or business life cycle. The case needs to be maintained and reassessed regularly to ensure that what is developing still makes sense; that is not to say that a project should stop half-way through because when reassessed it no longer makes sense, but that what is delivered or how it is delivered can change to provide a better fit.

16.2. Business cases for project delivery

When developing a business case for a new construction project, it is important to keep all options open as much as possible to avoid unnecessarily predefining solutions. Clearly there will always be a point in a project's development where decisions need to be made on what solutions to take forward (and these decisions will almost certainly lead to a degree of constraint on what other solutions can be incorporated), but these decisions should not be taken before a clear business case and clear requirements have been defined or before all relevant stakeholders have had the opportunity to provide input. It needs to be clear that the chosen solution is capable of meeting both the requirements and the business case (including any budget and programme constraints). A good example of this is the upgrade of Bank Station for London Underground (LUL), a very complex infrastructure project with a budget of £625 m. Instead of designing the project in detail before taking it out to tender, LUL carried out the usual technical studies, analysis of future passenger demands and flows and cost–benefit analysis, and shared these with the tenderers. The bidders were then asked to develop their own solutions to meet the specified requirements and achieve the highest benefit to cost ratio. Dragados won the tender, engaging with their supply chain to develop optimised solutions and, in addition to achieving a higher than anticipated benefit to cost ratio, the project was brought in £61 m under budget.

Early engagement of the supply chain undoubtedly offers the potential for business cases to be developed and tested prior to significant costs being incurred, and can provide a collaborative environment where detailed information from the supply chain can be fed into the process, thus providing greater certainty that the business case is valid. This can also be achieved by working collaboratively, or assessing business cases across a number of projects, not just one at a time. This provides certainty for the supply chain (which is particularly important when developing new technologies), and can enable risk to be spread over a number of projects, as well as learning and potentially technologies to be shared. It can be particularly beneficial when moving to a more industrialised construction approach as it can support economies of scale, and working with the same or similar teams while using the same technology is always likely to be beneficial.

One of the important factors in the Bank Station project was the availability of detailed technical information which provided comfort for the supply chain in developing their solutions. The availability of such information can also sometimes enable rapid development of business cases, which can be of significant benefit (for example, when looking to purchase land for developments). In 2018 PCSG and Wienerberger developed a tool to assess the potential viability of a residential development in minutes, and simultaneously assess the impact different construction technologies and methodologies would have on that viability. This was driven by the desire to illustrate the potential value of different manufactured solutions across a project, while providing cost certainty, accurate bills of materials and cashflow forecasts, combined with development-wide project viability assessments. The tool relied on accurate product data and defined construction technologies and processes, combined with local market and supply chain data to provide accurate cost, programme and benefit analysis. In a similar fashion, TestFit (described in more detail in Chapter 17) enables users to quickly generate virtual models of developments and assess dwelling mix and numbers, costs and programmes based on site constraints. As with the viability tool, this can rapidly provide valuable input into a business case for a development.

16.3. Business cases within the supply chain

To deliver transformational performance across a project, never mind a whole industry, requires all stakeholders to play a part. To enable that to happen, individual players need to be comfortable with their own business models and contractual boundaries to ensure that developed solutions are mutually beneficial (or at the very least nonconfrontational) in order to succeed. It is common for businesses who are looking to introduce new models to make small, incremental changes or additions to existing models rather than making significant shifts in one go. While it is understandable, this tweaking may no longer be enough to deliver sustainable new models that will take a business or industry to where they need to be in the years ahead. One approach is to create almost a portfolio of business models that can be used in different circumstances. For example, are products sold or are services hired (temporary classrooms, for example, instead of new buildings)? Are solutions supplied and installed by the same business or sold for others to install? Providing business model options can enable a business to significantly increase their potential pipeline by avoiding reliance on a single model that may restrict its addressable market. Such an approach can make a business more agile and capable of coping with shifting client or industry demands instead of being over-reliant on a business model that may no longer be valid. It enables new models to be tested while minimising disruption on other parts of the business.

As described in Section 16.2, it is beneficial where possible to offer the supply chain the potential to innovate or provide a range of solutions to requirements instead of merely asking them to respond to a tightly defined scope. In 2019 the Industrial Strategy Challenge Fund (ISCF) project 'Digitally Connected Supply Chains' (CPA, 2019) developed an approach to create a supply chain digital twin. This provided an up-to-date and detailed model of the supply chain for a project, and enabled supplier offers to be assessed, combined and aligned to provide a complete solution, with visibility of all contractual interfaces and potential gaps in offers. Such a model enables the supply chain to innovate and potentially enter new markets based on its capabilities, without restrictions being defined by traditional package boundaries. For example, a logistics company could in theory openly bid to provide transportation services for a number of suppliers across a project instead of the products of one supplier or one traditionally defined package alone, and this offer would then be assessed based on the potential value achieved against competing offers to deliver transportation services on a product-by-product basis. This not only benefits the individual businesses, but also the client can potentially receive additional, unintended benefits from their supply chain, and the case can be continually monitored throughout the delivery process.

Another challenge that is common in smaller businesses is the time and resources it takes to provide estimates for clients (often with little to no design information), and to have the confidence that those estimates will deliver them a profit. It is not uncommon for small builders or trades, for example, to spend the day working on delivering projects, and evenings or weekends pricing up new projects. If they over-estimate the project or take too long they may not get the work because the case may not work for the client, but if they under-estimate it they may not make their necessary returns to stay profitable. To address this, Sort It AI focus their business on enabling trades or those working on small projects to price their work in minutes. Described in more detail in Chapter 18, they do this by first enabling quantities to be calculated from images of drawings, and then by acting as a price and stock comparison website with local builders merchants. This enables the contractor to quickly provide an accurate and up-to-date estimate to the client, while also reducing the time and effort that needs to be invested in bidding for projects, so the benefit is felt on both sides of the line.

16.4.   Business cases and manufacturing-led construction

In moving to more manufacturing-led approaches to delivery, the relationships between different parties changes, as do the business cases of both project and individual businesses. The system provider is likely to take more of a leading role, which can change the dynamic from a more traditional project team and have an impact on the case for other suppliers and contractors. Where a business case identifies that a manufacturing-led solution be adopted, it is generally unwise for this to be the end goal in the same way that identifying any solution in an initial business case can be problematic. As business cases progress and become more defined, it may well be that a specific solution is identified and followed, but it should not be the starting point.

There is to some degree a clash in investment cycles between a typical construction manufacturer and the relatively short-term project focus of the construction industry. The potentially large investments in production facilities can require a five-plus year perspective, and that is a significant pipeline that needs to be filled to keep production at the desired rates to make the investment worthwhile. Imagine, for example, a business producing volumetric units for the residential market. A decent size of production facility may deliver 8 000 modules per year. If a house is made of four modules, that is equivalent to 2 000 houses per year from that one supplier, all of which need to be supplied at a steady rate to keep the facilities working optimally and to avoid having to provide additional warehouse facilities to store completed units. Developments, on the other hand, very rarely have a consistent rate of demand along the project cycle, and may face delays due to planning approvals or slow sales rates. Even when delivering to multiple projects, the demand is unlikely to be consistent unless it is greater than production capacity, in which case there either needs to be additional capabilities to meet the demand, or a method of managing client expectations where delays are likely. It may therefore be beneficial to have production facilities that can either deliver 2D panels or 3D volumetric units, which is likely to open up capacity for other markets and smooth the production curve. It also means that there may be the option to buy in certain elements when capacity is really pushed. Another alternative model is to not rely on a purpose-made production facility at all, but instead to use existing supply chains to deliver an offer, such as Modulous proposed (see Box 14.2).

Another aspect for a manufacturing business to consider is whether it should only supply units, whether it should work with an approved network of installers or whether it should install units itself. There are pros and cons associated with each model; it can be useful to have an element of control over installation to safeguard quality and to share knowledge of the good and bad of production and installation within a business, but it can also be dangerous to overstretch a business's capabilities if it is not comfortable with site activities. This is where there should be a very clear business case for a building manufacturer's operations to define where the boundaries of its offer sits, and to stick to that model or models and not allow scope to creep beyond available capabilities.

Initial business cases should also be open to innovative solutions that the client may not have considered – for example, leasing buildings or elements instead of buying outright. An example of leasing building elements is Jan Snel in The Netherlands. They lease steel frames for modular buildings, collecting them at the end of a rental period and refurbishing them for future use. They also deliver permanent buildings which, as described earlier, gives them the benefit of having more than one business model to meet client demand and reduce reliance on a single market.

16.5.   Analogue or digital business case processes

While there are certain aspects of business case development and testing that can be automated and carried out digitally, other activities are always likely to need human input. For example, the provision of financial modelling, demand, material availability and pricing and some elements of technical studies can be collated or produced automatically, but they are really inputs into a business case. There is still likely to be a need for human judgement and nuanced decision making on a final case.

Software tools do exist to assist in the development of business cases, such as Edison 365 and similar applications, but these still require human intervention. Once a business case has been developed, however, it is useful to have it in a digital format that can be checked and maintained at relevant intervals. Project management software tools can of course be used to track progress against programmes, but the business should still be reassessed at key stages of a project to ensure it is still on target to be delivered.

16.6.   Conclusion

In many ways, general best practice in business case development and management is also beneficial when it comes to manufacturing-led businesses or interventions; the need to clearly define desired benefits and requirements, and to not close off potential solutions too early in the process without good reason. Approaches to whole life value, such as the Five Case Model and Value Toolkit, also enable vastly different solutions to be compared based on the value that they deliver, not on preconceived ideas or historical practice alone. However, where new solutions are proposed there still needs to be supporting evidence that the solution is likely to be successful. To provide that supporting evidence is likely to involve collation of performance data, test cases, previous similar projects or solutions and guarantees and, while that evidence should be held digitally to ensure that it can be easily accessed and validated, it is likely to be prepared and collated manually in the first instance. However, the intention should always be to manage this supporting information in such a way that it can automatically be updated and shared to support future business cases.

Using the framework described in Chapter 6, Figure 16.1 identifies some of the automation technologies that can potentially be of use to support business casing activities. Their application is not likely to automate the decision-making process, but to provide supporting information for humans to consider. The assessment of business cases is likely to remain a manual process requiring human assessment and intuition. As a project or business develops over time, however, and a business case is revisited, it is typically a straightforward process to automatically update supporting information to enable reassessment by the relevant parties, but only where the information is sufficiently structured and accessible. For example, are programme and financial models automatically linked to business case tools? Making data accessible in this way can significantly reduce resource requirements in preparing and managing business cases, but there should always be a human check that all relevant data has been incorporated into a business case before next steps are decided.

Figure 16.1 Automation technologies relevant to business casing (Author's own)

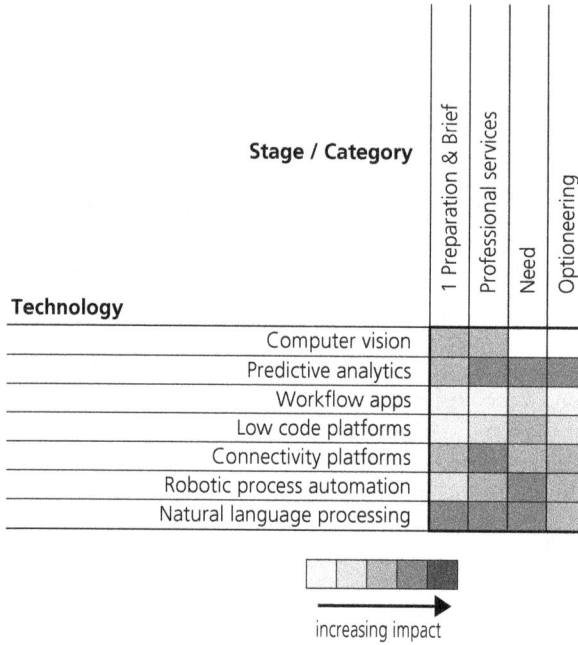

Technology	1 Preparation & Brief	Professional services	Need	Optioneering
Stage / Category				
Computer vision				
Predictive analytics				
Workflow apps				
Low code platforms				
Connectivity platforms				
Robotic process automation				
Natural language processing				

increasing impact

REFERENCE

APM (Association for Project Management) (2019) *APM Body of Knowledge*, 7th edn. APM, Princes Risborough, UK.

Construction Innovation Hub (2020) *Value Toolkit*. Construction Innovation Hub, UK. https://constructioninnovationhub.org.uk/our-projects-and-impact/value-toolkit (accessed 18/07/2024).

CPA (Construction Products Association) (2019) https://www.constructionproducts.org.uk/media/537687/supply-chain-digital-twin-release.pdf (accessed 2/8/2024).

HMT (Her Majesty's Treasury) (2022) *The Green Book – Central Government Guidance on Appraisal and Evaluation*. The Stationery Office, London, UK.

Steve Thompson
ISBN 978-1-83608-599-7
https://doi.org/10.1108/978-1-83608-598-020241020

Chapter 17

Design automation and manufacturing automation

17.1. Introduction

Over the last decade or so, there has been a lot of discussion on the likelihood that jobs in the construction industry may be automated in the not-too-distant future, and much of the focus in recent years has been on the automation of professional services, such as architects, engineers and cost consultants. While much of the activities carried out by these professions can be described as knowledge work, and so potentially harder to automate than physical or nonphysical repetitive tasks, the development of advanced AI systems and a move towards greater standardisation or industrialisation in construction potentially make it easier to automate such activities. However, there is still a long way to go before jobs will be replaced completely by automation technologies as, along with technical challenges involved in automating tasks carried out by those roles, there are issues such as risks, liabilities and cultural norms to deal with. In addition, obviously the roles of architects and engineers involve much more than design activities, and some of their additional work is potentially much harder to automate. Instead of being automated completely, it is more likely that roles will be augmented by technologies with the aim to improve the quality, consistency and volume of output that can be delivered.

To understand the technical feasibility and appropriateness of automating design tasks, it is useful to first break down what types of activities are likely to fall under the banner of design. For this purpose, the following four categories have been identified

- *creation* – the development of a design solution, from initial concept through to detailed design. Typically this is based on the experience, knowledge, judgement and skill of a designer and can involve a significant amount of creativity in addition to technical know-how. To some extent, experience and knowledge can be replicated by training AI systems on vast amounts of data from previous solutions and clearly defining sets of rules, but that is not a replacement for creativity and judgement; an AI does not understand why a certain solution is or is not appropriate, rather it presents solutions that meet defined criteria based on previous examples. However, as Bernstein (2022) suggests, while a computer cannot reason like us, it may still be capable of carrying out tasks that would require judgement when done by a human. While they may never be creative in the same way as humans, they can deliver results that humans may interpret as being creative – it may just be that it does this in a completely different way from us
- *presentation* – the process of describing the design to others, possibly through drawings and specifications. This is potentially the easiest of the four categories to automate, especially as today it largely (but not always) involves presenting information that has been created using digital tools

- *analysis* – determining how a solution is likely to perform under certain conditions, what it is made of (for example, a bill of materials), and potentially how long it will take to deliver and how much it will cost
- *sharing* – enabling and coordinating collaboration with other disciplines, and publishing information either for collaboration, manufacturing or construction.

The following sections will cover these categories at different stages of design development, through to production of manufacturing-led solutions which are likely to benefit from consistent, accurate and reliable design data as inputs for production.

17.2.   Early-stage design development

The early stages of design development can be a lengthy process; interpreting site and planning constraints, developing and calculating whether all accommodation can physically fit and what that will mean for massing, judging the potential quality of internal spaces based on orientation and access, estimating costs and, for residential developments, understanding the potential mix of units. Before discussing the potential implications of automation on design development, it is worth briefly exploring site selection, planning and local contextual data that feeds into the design process and can take many weeks to collect manually. Tools such as LandHawk enable users to quickly search for available land and to understand existing ownership and potential numbers of residential units that can be delivered on that site. Addland identifies local land for sale, but also provides easy and instant access to detailed information on any historical planning applications relating to a site as well as any planning, conservation area or other constraints that apply. These tools can significantly speed up site research and analysis and provide a good baseline for development.

As discussed in earlier chapters, the move to manufacturing-led approaches, and volumetric approaches in particular, can be challenging without a relatively smooth pipeline of projects to keep production running efficiently, and to finance the production of units prior to receiving payment. Planning is often cited as providing a challenge to this, firstly because of potential delays, but also site-specific constraints that may impede standardisation. The New South Wales Government are developing an interesting approach to solving this by exploring a 'pattern book' approach. The intention is to have a series of developed designs, which are published and known to be suitable for prefabrication, and if a scheme follows the published plans, planning will be preapproved as long as it meets site-specific constraints. This approach also means that suppliers have a good understanding of what is likely to be required for these projects and can prepare accordingly.

Before moving on to look at automation and generative design, it is worth explaining the difference between the two. Automation involves using computers to undertake repetitive tasks quickly and efficiently (such as collating and presenting data, or adding floors to a building based on an input parameter) or carrying out specific design processes such as calculations. Generative design goes beyond this, and uses complex algorithms to generate a number of design solutions based on input parameters. Such an approach can enable a large number of solutions to be generated very quickly and can lead to innovative solutions that may not have been considered traditionally (in some ways delivering creativity, but likely to be among a wider field of more predictable solutions). In the language used earlier in this section, generative design is used to create, whereas automation is more commonly used to present, analyse or share.

Once a site has been selected and analysed, a next step may be initial massing to understand how buildings will sit within a site context. Autodesk Forma (illustrated in Figure 17.1) can be used to quickly model development sites and analyse the results in terms of areas and volumes and the impact on local environments. It automates the development of buildings based on inputs, such as building typologies and numbers of floors, and can then automate analysis to see how the resulting buildings are likely to impact on access to daylight, noise and wind. The models can then be exported to other software packages for further development.

Developing an early-stage design, even an initial concept, can require plenty of experience in order to potentially solve a multitude of challenges. This does not mean that every issue must be fully resolved straight away, but there needs to be a level of confidence that the developing concept can be realised in some way once further developed. With a move to manufacturing-led construction or industrialisation, that confidence can often be provided by knowing the detailed capabilities and restrictions of a building system and ensuring that these are fed into the design process. However, beyond the form of construction, there can also be other ways in which to recreate and in some ways automate the application of knowledge to solve complex challenges.

When considered carefully, there are likely to be a number of characteristics or possibly rules that apply to projects of a similar nature, and rules can enable automation. For example, in residential developments there are certain space requirements that are commonplace in different sectors, such

Figure 17.1 Autodesk Forma is used to generate, develop and analyse initial site massing (Source: Autodesk Forma)

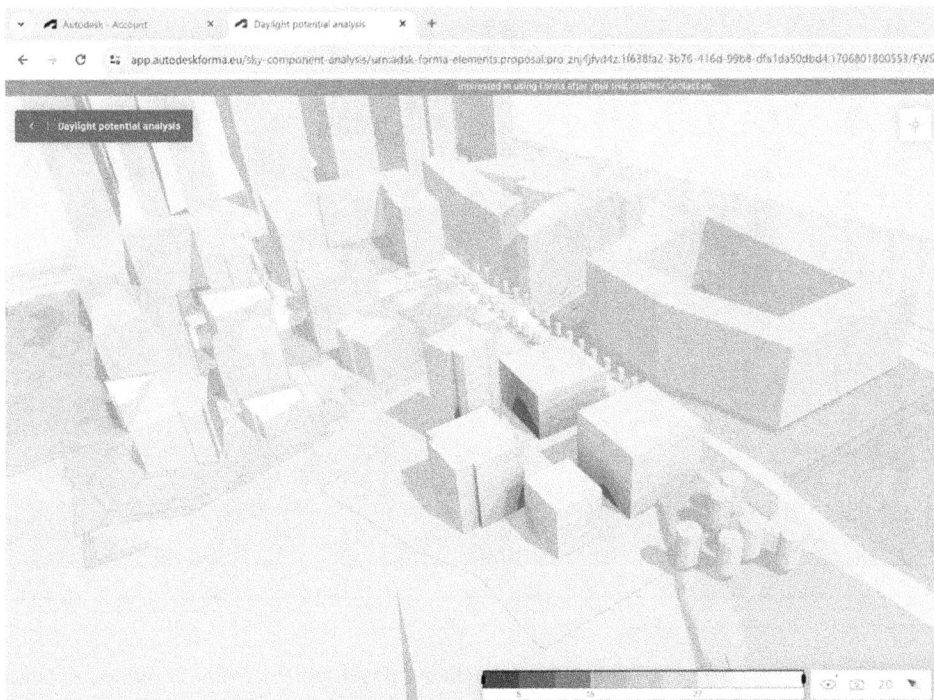

as bathrooms are likely to be of a similar size and with a limited number of layouts, and bedrooms will usually have similar layouts and requirements in a given subsector. This can lead to a preferred set of house or apartment types for a developer and, once apartments or houses have been defined, their application across a development can largely be automated. In the residential sector, TestFit is an example of a generative tool that quickly generates solutions based on user input to test the viability of new developments.

Box 17.1 TestFit

TestFit is a feasibility-stage generative tool that uses real-time artificial intelligence to rapidly develop and test solutions, enabling users to quickly assess the viability of a particular development site. Based on user input, such as target dwelling mix, parking ratios and any necessary set-backs from site boundaries, TestFit develops the best deliverable solution. If the site is in the USA, users can select a particular land parcel as a starting point, otherwise they can draw the site boundary manually. They then select a core building configuration (such as podium, tower, etc.) and TestFit creates a solution, which can be dragged or edited in real time to illustrate the implications of any changes. At the same time as the model is adjusted, costs, quantities and unit ratios are also updated, enabling users to quickly assess the viability of a proposed development. TestFit can currently be used for residential, mixed use or industrial units. Figure 17.2 illustrates results from TestFit.

Figure 17.2 TestFit uses real-time artificial intelligence to generate, develop and test solutions (Source: TestFit)

Skema is another interesting solution to rapid scheme development, with the added strength of being able to learn from a practice's previous successful projects (which they describe as a BIM knowledge reuse engine). Using machine learning, Skema creates concept catalogues from previous BIM models and uses these as part of the generative solution for new projects.

Not all sectors necessarily have the same levels of typology definition as residential, but that does not restrict the use of automated or generative models applied to other sectors. An example in the highways sector is the rapid engineering model (REM) described in Box 17.2. In highways design there are many clearly defined design rules when considering the safety and efficiency of a highway, and these form an ideal basis for automation of design.

Box 17.2   Rapid engineering model (REM)

The REM was developed by Bryden Wood for Highways England to automate the design of smart motorways, with the notional aim to be able to design a motorway in a day. In simple terms, the developed technology is an advanced configurator, taking the many published rules associated with designing a motorway, the know-how of experienced highways engineers and a range of product and system specifications, and hard coding them to enable them to be automatically applied to a given context and set of requirements. That in itself is no small task. The technology takes in a range of inputs and optimises layouts accordingly on a project-by-project basis to provide 3D models and technical specifications of a complete scheme. It also highlights where programmed rules are likely to be broken to enable human consideration.

Given a (typically very complex) set of geometric and environmental inputs and design constraints, the REM technology generates solutions that meet defined requirements, including pulling in the predesigned solutions for gantries and other structures.

Generative design can also be very valuable in the design of gradients, as illustrated in tools such as Bentley's OpenSite or OpenRoads. Whereas traditionally often the first compliant grading solution would be selected and progressed due to time and budget constraints, using these tools that work with user-defined constraints (such as allowable elevations and slopes), a series of iterations can be progressed to generate the optimal design solution. Such technologies can save significant amounts of time while also minimising the amount of material wastage on site.

Generative design capabilities are now in some cases becoming available within mainstream design tools, such as AutoDesk Revit. They provide the ability to rapidly model and easily compare a large number of solutions to a given challenge, but they do typically require significant training and practice to operate effectively on anything but the simplest of tasks, so are not necessarily readily accessible to the novice or occasional user in the way that tools such as Forma and TestFit are. In addition, it is important to minimise hard coding rules where possible, to provide flexibility and prevent over-constraining solutions. Machine learning is likely to have an impact in the future in helping to define constraints, hopefully reducing the need for systems to be hard-coded.

17.3.   Design development

In principle, key stages of design development are traditionally linear, with iterative processes within a stage. As a project progresses from one stage of development to the next, the level of detail needed increases, as does the complexity. However, when moving towards a more manufacturing-led approach, it is likely to mean that more detailed information is available earlier in the development process. While it is generally not advisable to include detailed information before it has been agreed on (for example, before a solution has been chosen), when a specific technology has been selected and confirmed, having more information earlier in the process can provide confidence and certainty in what can be delivered. Figure 17.3 presents four basic models for when detailed product information is applied to take a proposal from initial to detailed design and beyond. In a traditional development, specific products or more refined generic products are likely to be manually applied to add necessary detail to take a project forward. The majority of products are likely to be applied at this stage to eventually prepare the scheme to be realised in a construction process.

In model (b) of Figure 17.3, the initial design is developed as normal, but then the designer can automate the application of multiple manufacturing-led products to that initial design to take it forward. KOPE is a great example of this and is described in Box 17.3. Such an approach provides

Figure 17.3 Design development automation models (Author's own)

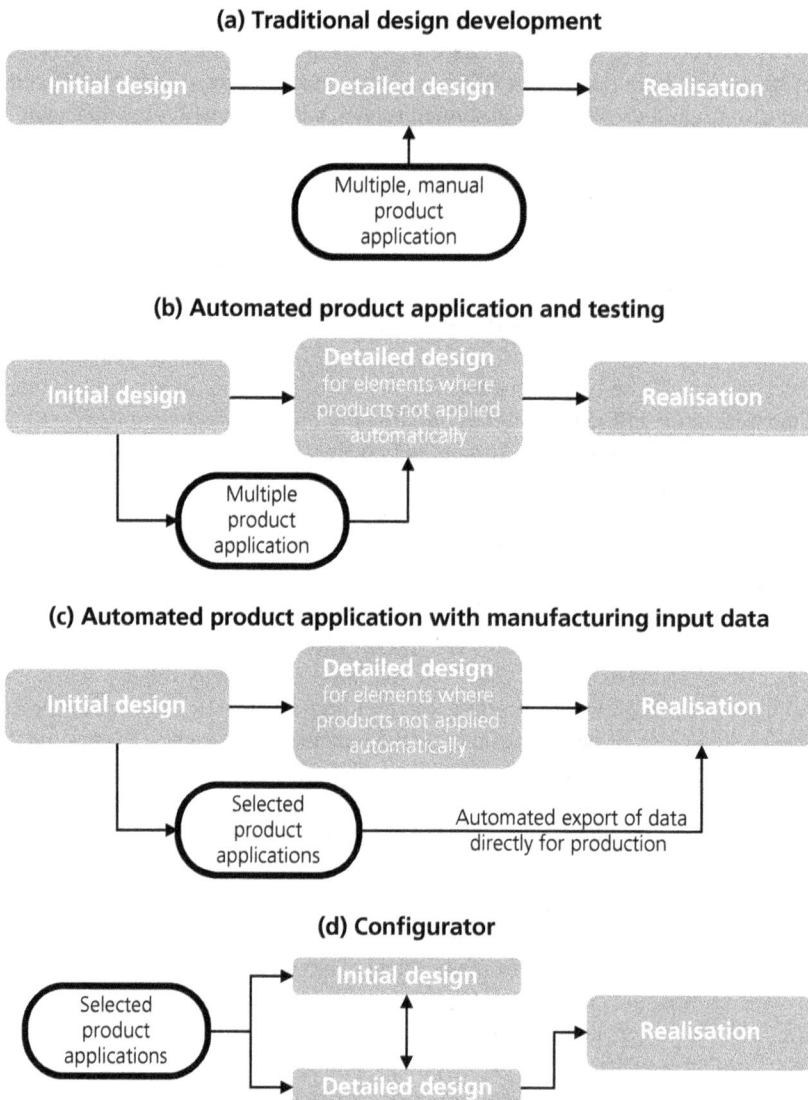

(a) Traditional design development

Initial design → Detailed design → Realisation

Multiple, manual product application

(b) Automated product application and testing

Initial design → Detailed design for elements where products not applied automatically → Realisation

Multiple product application

(c) Automated product application with manufacturing input data

Initial design → Detailed design for elements where products not applied automatically → Realisation

Selected product applications

Automated export of data directly for production

(d) Configurator

Selected product applications

Initial design ↕ Detailed design → Realisation

Box 17.3 KOPE

KOPE enables users to automate the application of real-world products to a project. Starting with an IFC (industry foundation classes) model of a scheme, users can select which products to apply from those that have been preconfigured on the platform, and which elements they apply to (for example, which framing system to apply to a wall type). KOPE then quickly and automatically applies the products to the model in the most optimum solution possible, complying with any rules identified by the manufacturer. This ensures that the solution is constructable with the selected products and systems and provides supporting information on quantities which can then be shared directly with the manufacturer and others, along with layout drawings. This quickly takes an early-stage design model through to a detailed model and enables different product solutions to be quickly compared and tested. Multiple products can be applied simultaneously (for example, wall and floor systems). Once products have been applied to the model, changes can then be made by editing parameters (for example, stud centres). As a result of the level of detail incorporated, costs can also be estimated, and any areas where a product's application may require a more bespoke solution (for example, where a curved wall is required) are clearly identified. Figure 17.4 illustrates KOPE in action.

Figure 17.4 KOPE automates the application of real-world products and systems to projects (Source: KOPE)

As a result of using the platform, KOPE identify a potential 75% reduction in time to respond to tenders and a 90% increase in accuracy of tender pricing responses. While products are directly accessible and can be explored within KOPE, it is supported by KOPE Market, where a vast range of products and systems can be searched for use in the UK and USA.

confidence that a solution is deliverable but also enables early comparisons of different systems and methodologies and provides accurate cost estimates for the elements where products have been applied. While such technologies do not create new designs, they develop initial designs as well as present, share and analyse the results.

Similar in some regards is the process presented in model (c) of Figure 17.3. In this case, it is the manufacturer or system provider who automates the application of their product or products to an initial design model, and they do so for different reasons. In this case, the application of products is done to provide accurate quantities, estimates and production information to the manufacturer initially, who can then verify that their system is deliverable for the scheme. From this model, they have data in the right format ready for production and can also check issues, such as availability and lead times, and provide accurate time and cost quotes back to the project team. When the time is right, manufacturing code can be exported from these models directly to the production line, thus directly influencing the realisation of the project, whereas in model (b) of Figure 17.3 this step would be carried out in a more traditional process. This effectively eliminates the need for detailed design of the elements that have been applied and goes directly to production. Box 17.4 describes Auto/Mate, a platform that delivers this model.

Finally, manufacturing-led technologies can be applied to designs through the use of a configurator. Configurators are often perceived slightly negatively in the design world as limiting design options, but it completely depends on their purpose; configurators can be rather complex and provide significant design choice and, for certain applications, are ideal for speeding up the design process and providing certainty. Configurators are used to ensure that the characteristics and capabilities of products and systems are fully considered in the development of a design, instead of retrospectively fitting products to an initial design. A configurator can be used to create a new design (with a defined set of components), present, share and potentially analyse a scheme. A perfect example of the proper use of a configurator is the Laing O'Rourke Bridge Configurator. The tool is used to configure a modular bridge system consisting of a standard range of precast concrete products that can be used to deliver single-span integral bridges, and configures all bridge elements including parapets, deck, abutments, wingwalls and piles.

While generative design tools in the built environment are typically restricted to the initial design stages, there are exceptions. RST (described in Box 17.5) is such a technology and has the ability to generate fully resolved solutions from scratch. Unlike a typical configurator that is limited to a small number of products, RST is effectively unconstrained and can simultaneously generate complex forms at the same time as dealing with detailed product application and interfaces. Generative design in the built environment is much more complex than in most sectors for anything other than form finding, and the reason is the need to manage a hierarchy of operations in such a way as to achieve the desired goal. It is this complexity that RST looks to resolve, and the method used is described in detail by ArchiConDes (Ravnikar *et al.*, 2015).

Augmented design is the concept of carrying out live analysis alongside design development (in a similar way to TestFit and KOPE in their different ways assess costs and quantities). However, in the future, augmented design is likely to go further and look at the performance of designs. Today, some degree of analysis can be quickly carried out directly within a modelling environment (such as daylight analysis in AutoDesk Forma), but doing this with a detailed design model is complex and typically requires exporting models for analysis. Chapter 9 discusses how this can be achieved in more detail.

Box 17.4 Auto/Mate by DataForm Lab

Auto/Mate differs from other product application platforms in that its focus is simultane-ously on the manufacturing and design perspective, as it looks to automate the design to production workflow. Starting with a design 2D or 3D model, a manufacturer applies their preconfigured product or system to a project (for example, floor cassettes applied to all suitable floors within a scheme) to create an optimised solution almost instantly. Armed with these results, the manufacturer can plan factory production in the most efficient man-ner, as well as provide accurate cost and time quotes, confident that the scheme can be delivered.

As part of the onboarding process for manufacturers, not only are the product's or system's capabilities and configuration rules developed but also the production machinery data for-mats and requirements are identified. As a result, when a product is applied to a project model, production code can be exported directly from that model to the machines that will later produce the finished product. DataForm Lab work with the manufacturer to identify current production methodologies and the potential for, and implications of, manufacturing process automation, as described in Box 5.1 in Chapter 5. Figure 17.5 presents the principles of Auto/Mate.

Figure 17.5 Auto/Mate by DataForm Lab to automate design to manufacturing for off-site construction (Source: DataForm Lab)

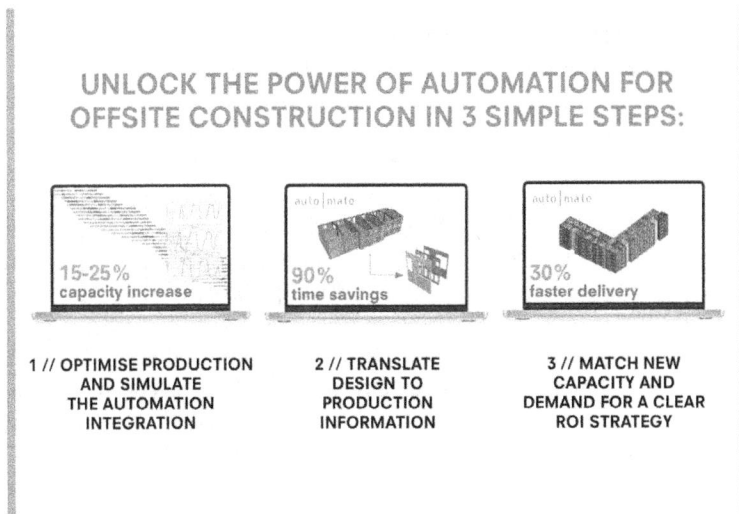

UNLOCK THE POWER OF AUTOMATION FOR OFFSITE CONSTRUCTION IN 3 SIMPLE STEPS:

15–25% capacity increase

90% time savings

30% faster delivery

1 // OPTIMISE PRODUCTION AND SIMULATE THE AUTOMATION INTEGRATION

2 // TRANSLATE DESIGN TO PRODUCTION INFORMATION

3 // MATCH NEW CAPACITY AND DEMAND FOR A CLEAR ROI STRATEGY

So, now that different types of design automation, augmentation and generative design have been described, what is the reason for using such approaches? Clearly one reason is speed – not to be able to finish a design quicker and just sit back and wait for responses from other project partners, but to be able to respond to clients' and other partners' requests in a more timely fashion, to be able

Box 17.5 RST by ArchiConDes

Figure 17.6 RST by ArchiConDes (UK) generates fully resolved, detailed designs from scratch, based on input parameters and either generic or specific products (Source: ArchiConDes (UK))

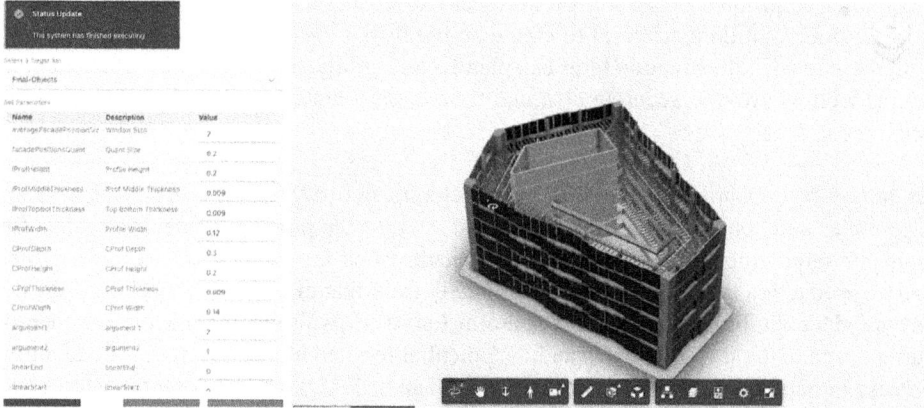

RST (Ravnikar Soper Technology) is an advanced generative design tool currently under development which produces fully resolved solutions from scratch. This means that it can generate unconstrained, fully detailed solutions based on input parameters and either generic or specific products, so creates initial designs that in theory do not require any more development and are construction or production ready. However, while the capability is there to take a single step from initial design straight through to production, it is more likely that fully detailed models will be assessed and refined by design teams and clients in a number of iterations, but with the confidence that whatever is generated is deliverable and could be taken straight to production. In a traditional design process, detail is only applied once a concept has been set, meaning that it can be difficult at a later stage of development to go back and make significant changes. With RST, the detail and concept can be regenerated and resolved simultaneously, providing greater flexibility in achieving the desired results.

Whereas most generative design tools are restricted to form finding at the early stages of design or retrieving and applying preconfigured solutions, RST has been developed to cope with the complexity of fully resolved designs without the need to overly constrain or simplify a product or asset's requirements. RST manages complexity by controlling how it grows through continually upgrading a model as it develops, based on immediately preceding operations without the need for simplification. It then focuses on the localised dependencies impacting components and their application. In simple terms, this means that a solution starts off with the simplest form and step-by-step features are added to create a more complete whole. It does this by constantly recalculating operations instead of adding detail onto previously resolved and fixed operations, which can lead to decisions being made without full visibility of the context and result in undeliverable or suboptimal solutions. Think of the way a tree grows in a confined space; the first branches are likely to grow to get as much light as possible, but that may mean that as branches grow off of these, the structure becomes overcrowded and stops some of the lower branches from developing. In a tree modelled in RST, the final solution would be modelled to optimise the whole, so it is similar to using sticks and mesh to initially train the growth to the desired form.

RST (illustrated in Figure 17.6) also automatically generates a dependency graph of relationships between components on the fly, whereas in most generative design tools this needs to be explicitly defined by the user. Initial analysis by ArchiConDes over a number of projects suggests that cost savings in the order of 14% and time savings of over 30% can be achieved by using RST (including an up to 90% time saving on detailed design and estimation).

to make design changes more efficiently and to be able to increase productivity. As discussed in Chapter 1, the construction industry needs to deliver more assets at a quicker pace than ever before, and so increasing the speed at which assets can be designed is a key part of that.

Beyond speed, reliability is a key driver. There are two aspects of reliability that are important here, the first being consistency. Automation has the potential to deliver accurate and consistent results, and thus minimise the potential for human error, but it also supports the potential for scalability across numerous developments. Secondly, automation can lead to fewer errors in that the solutions developed should always comply with the rules that have been defined or, at the very least, identify areas that do not comply. The Get It Right Initiative (GIRI) define an error as 'any action or inaction that results in a requirement for rework, a requirement for extra work, or produces a defect' (GIRI, 2023). In their research, GIRI identified 17 root causes of error in construction; the top three were inadequate planning, late design changes and poorly communicated design information, so all relating to the design stages and related activities. Automation can significantly reduce the likelihood of those root causes occurring if applied correctly in creating, analysing, presenting and sharing design information. Another key aspect of this is minimising the potential for clashes later in the design process or, even worse, during production or integration. Again, automating product application and clash detection can be carried out earlier in the process when moving towards a manufacturing-led or industrialised solution.

When automating design tasks, it will not always be the case that it is quicker than traditional approaches to develop a design (although it is always likely to be quicker to produce drawings and information on the design that has been developed). As has been discussed in this chapter, while some automation tools carry out their tasks almost instantly, others require potentially significant hardcoding by experienced users to deliver results. Even apart from the potential work involved in coding requirements or training artificial intelligence that some tools rely on, designers still have a significant role to play in defining the right parameters and selecting the right solutions from a potentially vast number of options. It is likely that for the foreseeable future, the user of design software will be held responsible for the results generated, not the provider of the software. As a result, there still needs to be real care taken in the design stages.

As buildings and other assets become increasingly more complex, it is more important than ever that software used to assist in design development can cope with that complexity in order to give confidence that developed solutions are likely to perform as required, and that is where analysis software is invaluable in assessing, predicting or presenting results.

17.4. Design to manufacturing

In a traditional work flow, the manufacturers of construction products or building systems are not typically engaged until the detailed design stage, where many of the key decisions on a project's design have already been made. There can then be a number of iterations between design and manufacturing to ensure that what is designed can be delivered by the selected manufacturers. Clearly, with a more manufacturing-led approach, the potential exists to engage earlier in the design process and hopefully ensure that both project and production can be optimised as a result. In doing so, it is important to not only consider the geometrical and performance implications of an identified technology but also the practical implications of how and when the solution is produced, assembled and installed, and the capabilities of the supply chain to deliver such a solution.

There is often a misconception that all production facilities are highly automated, which is not the case. Some construction product manufacturing facilities are still rather analogue in their approach, relying on machinery that has been operating for several decades and requiring manual instructions to be inputted into them. For these facilities or static assembly, what is really needed is information to be presented in a clear and consistent manner to increase the likelihood that instructions are inputted and carried out effectively. Where more modern equipment is available, it can be that design information is fed directly from a design model through to the production equipment by way of either a platform such as Auto/Mate or through modelling in software such as Vertex DB for framing solutions. Whichever approach is used, there needs to be a clear understanding of the information requirements for efficient production prior to spending significant time and resources creating information in a format that is neither necessary or useful. For this reason, early engagement with manufacturers can be invaluable, and any potential production constraints can be clearly understood. This is where platforms such as KOPE can be extremely useful as a starting point.

In early 2024, Autodesk launched Autodesk Informed Design, which enables manufacturers to share their product options with project design teams, including any product variation options and constraints and effectively acts as a configurator for products. This is another example of closing the gap between design and production, and promises to be a useful tool in supporting designers when moving towards manufacturing-led or industrialised solutions. More broadly, a design for manufacture and assembly (DFMA) approach can be utilised to ensure deliverability is considered throughout the design development process. However, DFMA is not enough; future operation, maintenance and replacement of systems also need to be considered. If a system is designed effectively to be manufactured and installed, but future accessibility for maintenance or repair is not considered, its effectiveness over the life cycle of an asset can be reduced, leading to future abortive work. With the growing use of 3D models and other data being handed over to clients on completion, those responsible for managing assets potentially have more information than they have ever had before, which is even more important when building systems are used which trades may not be familiar with when carrying out future maintenance or remodelling.

One of the key aspects of a DFMA activity is to look to design out waste, and this should not only consider the building system itself but also any other product or process that is impacted by such a technology. Designing out waste can include reducing the number of components, reducing the interfaces between systems and trades, reducing the number of processes required to assemble or install a system and ensuring that solutions fall within the capabilities of the production and delivery teams. The Construction Innovation Hub developed the Construction Product Quality Planning (CPQP) process to help in the development and application of new products into construction, but the process does not directly cover site-based assembly processes and is focused on applying products at scale. A more simple and flexible approach to assist in the optimisation of a solution for smaller scale projects in particular is the Five Ps model. The simple model enables users to consider the implications of new products or processes and where they can be optimised, and can be applied to any stage of the application of a solution.

17.5. Model quality and checking

Many of the tools used to automate parts of the design-to-manufacture process require design models as inputs and, while BIM and 3D modelling have been around for quite some time, the quality and reliability of design models still varies significantly. As with many things, poor input data can lead to poor or limited results, so it is important to ensure models are of sufficient quality to be able to identify what different elements are and to be able to take off relevant quantities, for example. Tools such as Simplebim are easy to use and effective for such activities, as are Solibri Model Checker and Autodesk Model Checker.

Whereas all these tools can be used to check compliance with information requirements, Solibri can also be used to check compliance of the design intent against defined rules. In the USA, software such as Upcodes takes this a step further and, based on a project's location, can be used to check compliance against local requirements without the need to manually define individual rules.

The ability to check compliance against regulations depends largely on how the regulations are written in the first place. In the UK, for example, many aspects of the Approved Documents are not currently written in a way that can be automatically checked for compliance, as they do not always identify clear pass or fail criteria, so are subjective instead of objective requirements. Technologies and methodologies do exist internationally, however – for example, CORENET e-PlanCheck is a web-based service used in Singapore to check compliance using IFC models and covers specific aspects of architecture and building services.

Clearly, models will not exist for all projects, especially smaller scale or possibly domestic models, and for these circumstances the potential for automation of design-to-production processes is significantly impacted. However, the key in those circumstances is still clear communication, and it simply means that the stage of the process where design data is input into the design-to-manufacturing process is postponed. Software such as Sort It AI use computer vision to be able to take images of drawings and produce schedules of elements, which can then be fed into the procurement process, but do not go as far as producing digital content that can be used for compliance checking.

17.6. Technology relevance summary

Using the modelling framework described in Chapter 6, Figure 17.7 summarises the potential relevance of different automation technologies to the design and manufacturing aspects of construction activities.

As this chapter has illustrated, solutions do exist to support the automation of design creation, presentation, analysis and sharing, and these can be used to link design with production capabilities. The relevance of such technologies largely comes down to the capabilities of design and production teams and facilities, and the stage at which engagement between the two can occur. The move towards manufacturing-led or industrialised construction can and should significantly improve this link and, as a result, open the door to automation of design-to-manufacturing activities to augment existing capabilities, but it is still likely in the short to medium term that ultimately design decisions and responsibilities will still remain with the design team.

Taking a design from concept to detailed design either needs to follow traditional design approaches or utilise platforms such as KOPE or Auto/Mate to act as the link between design and manufacturing. An alternative in the future will be technologies such as RST that can develop fully resolved and unconstrained designs from the start of the design process, but human decision making and inputs will still remain important. As is discussed throughout this book, it is unlikely that entire roles will be replaced by automation technologies in the short term, more that human capabilities will be augmented by technologies that enable greater quantity and quality of output with the same resources.

To finish this chapter, it is important to recognise that design is not only about providing information for production. The information that is produced at the design stage is also key to the pricing of construction works, procurement and forward planning for the operation and maintenance of assets. The information produced at this stage must provide useful input for these activities, and so any technologies used to automate design activities must also be capable of smoothing the interface with such work. The following chapters will discuss this in more detail.

Figure 17.7 Technologies relevant to design automation (Author's own)

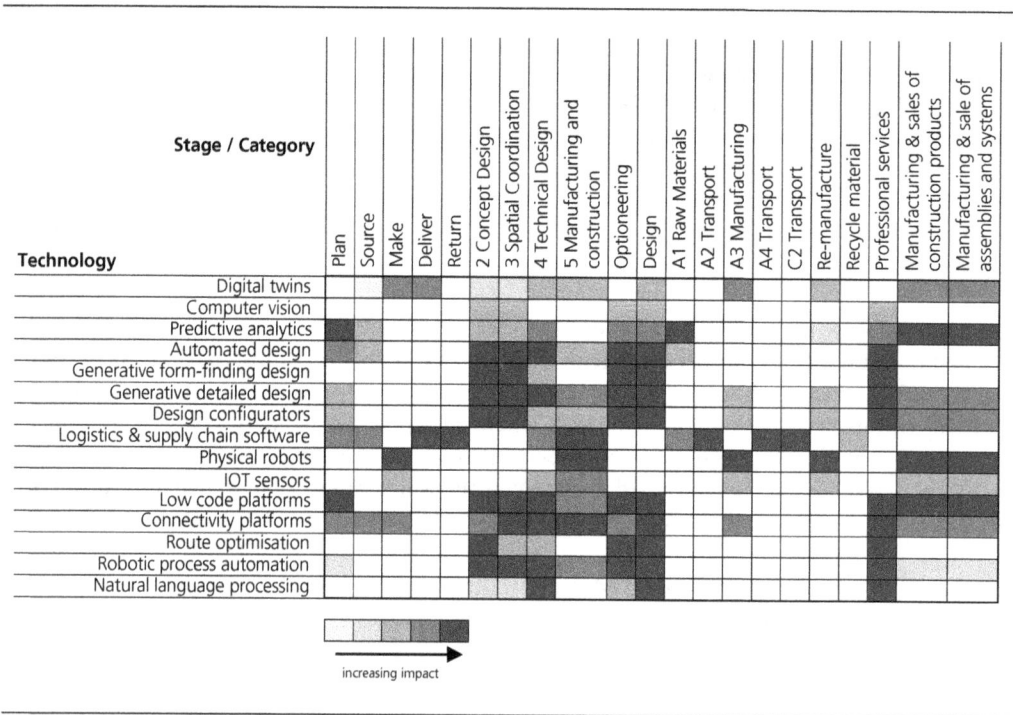

Technology \ Stage / Category	Plan	Source	Make	Deliver	Return	2 Concept Design	3 Spatial Coordination	4 Technical Design	5 Manufacturing and construction	Optioneering	Design	A1 Raw Materials	A2 Transport	A3 Manufacturing	A4 Transport	C2 Transport	Re-manufacture	Recycle material	Professional services	Manufacturing & sales of construction products	Manufacturing & sale of assemblies and systems
Digital twins																					
Computer vision																					
Predictive analytics																					
Automated design																					
Generative form-finding design																					
Generative detailed design																					
Design configurators																					
Logistics & supply chain software																					
Physical robots																					
IOT sensors																					
Low code platforms																					
Connectivity platforms																					
Route optimisation																					
Robotic process automation																					
Natural language processing																					

increasing impact

REFERENCES

Bernstein P (2022) *Machine Learning: Architecture in the Age of Artificial Intelligence*. RIBA Publishing, London, UK.

GIRI (Get It Right Initiative) (2023) *The Use of Technology to Reduce Errors in Design and Construction*. GIRI, London, UK.

Ravnikar E, Ravnikar M, Luzar B and Soper R (2015) *The Way to Efficient Management of Complex Engineering Design, Towards Solving the Social Science Challenges with Computing Methods*. PL Academic Research, Frankfurt, Germany.

emerald PUBLISHING   ice

Steve Thompson
ISBN 978-1-83608-599-7
https://doi.org/10.1108/978-1-83608-598-020241021

Chapter 18
Supply chain management and procurement automation

18.1.   Introduction

This book describes some of the many challenges facing the construction industry and wider built environment sectors, and the different technologies that, if adopted, are likely to help in meeting those challenges. It is important to recognise, however, that in many ways procurement sets the tone of interventions and can dictate whether new technologies are applied or whether projects are delivered as they always have been in the past. Manufacturing-led construction and automation technologies must be led by modern procurement practices and contract management, and outcomes-based approaches that are built on trust, respect and collaboration. In other words, if traditional procurement approaches are used, there is a good chance that the same results will be delivered. Procurement processes can limit the adoption of new technologies and approaches where there is no room or incentive to change and limited opportunities to take learnings forward to future interventions.

Automation and manufacturing-led construction, while certainly beneficial to individual interventions, are likely to be most valuable when considered over a number of projects to benefit from standardisation, repeatability and shared investment costs. At the same time, the decision to use such technologies on a project is best taken early in its development. To enable that to happen, there either needs to be early supply chain involvement (ESI) or projects need to be considered as part of a wider programme or portfolio. It is important to note the difference between early contractor involvement (ECI) and ESI. ESI can involve organisations of many types and tiers, such as product suppliers, specialist contractors and consultants as well as the more traditional integrators or Tier 1 contractors, whereas ECI typically only involves Tier 1 contractors. As described in Chapter 3, the majority of work on a construction project is often carried out by subcontractors at lower tiers of the supply chain, and so it is important to have their involvement early in a project to make sure that the potential to deliver the desired outcomes is maximised. A contract such as the Framework Alliance Contract FAC-1 is set up to support multiparty collaboration, and is suited to including organisations of any size, with the right capability. Such approaches enable lessons to be learned and shared across multiple projects and for objectives, success measures, targets and incentives to be aligned, thus reducing risk. It also provides the ability to ensure that risk is owned by the party or parties most able to manage it, not just passed through different tiers of the supply chain. NEC also provide contract options to enable ECI.

18.2. Procuring for value

Procurement needs to enable outcomes to be procured and delivered with consideration for the whole life of an asset, not the delivery phase alone. Good outcomes are unlikely to be achieved unless procurement is planned to enable them.

A fundamental principle that is often overlooked in the construction industry is that contracts should be profitable for all those involved. Fair returns and expectations need to be reasonable for suppliers at all levels of the supply chain if they are to remain focused on delivering the best outcomes for the client and for the market to be sustainable. The *Construction Playbook* (HMG 2020) identifies that the traditional construction supply chain can at times be very opaque, and it can be difficult to see or understand the relationships between different parties within the chain. That is an issue for payment practices and valuing of work carried out at lower tiers, but also in limiting the potential those at lower tiers have to positively influence the outcomes of an intervention. As described in Chapter 3, the majority of work in the construction industry is actually carried out by these lower tiers, and if those operating there are being pushed to deliver at the lowest cost and without the potential to offer alternative solutions, then innovation and increased value are unlikely to be delivered. For these reasons, it is important that an outcome-based approach to procurement does not stop at the appointment of a Tier 1 integrator or supplier; the approach should filter through the supply chain to have a real impact. If it does not, organisations selected on the basis of lowest cost alone will actually be those that are least incentivised to deliver a project. Procuring based on value must be beneficial to all parties if it is to succeed; simply asking organisations to rebid work at lower prices is not sustainable and can lead to them looking for other means to increase their income, such as through claims.

There is always a danger that a client's requirements change or are insufficiently defined in the first place, and changes are more likely to be adopted efficiently in a collaborative environment with a focus on shared values than in a confrontational arrangement where organisations are already operating on minimal margins and looking for opportunities to make a better return. Evidence shows that projects delivered in a collaborative environment are 15% more likely to be completed on time and 44% more likely to be within budget (Constructing Excellence, 2011). So, how can the construction industry move from a confrontational, short-term project model to a more collaborative, mutually beneficial and whole life model?

The objectives of collaboration can be difficult to translate into procurement and contracting, and can result in terms such as 'good faith' being used rather than defined commitments that need to be met. The FAC-1 contract described in Box 18.1 is a multiparty framework contract that identifies shared objectives, targets and incentives and enables a collaborative agreement between all alliance members (which can include suppliers and subcontractors).

Frameworks typically offer the benefit of a potential pipeline of work and the ability to take learning from one project onto others, but not all frameworks are successful in delivering this and can still be focused on delivery at the lowest capital cost instead of whole life value. In addition, they are not always accessible to smaller businesses.

In 2020 the Department for Levelling Up, Housing and Communities (DLUHC) commissioned a review of construction frameworks (Mosey, 2021) with the purpose of developing a gold standard for frameworks that would support the aims of the *Construction Playbook* (HMG, 2020), including a focus on delivering value. At the time, there were over 2000 active public sector

Box 18.1 FAC-1 contract

FAC-1 (Alliance Forms, 2024) is a framework alliance contract, a multiparty umbrella contract that is designed to integrate the activities of consultants, integrators, subcontractors, suppliers and others of any size and align their interests with those of the client. It is an umbrella contract which sits above other contracts and agreements, and defines relationships and processes not covered by those contracts.

FAC-1 defines clear objectives, success measures and targets, which are then linked to agreed incentives for alliance members, such as the potential for further work through the alliance. As a multiparty agreement it provides opportunities for SMEs and subcontractors to become alliance members where their capabilities are aligned with those of the client, and therefore can operate at any tier of the supply chain. It promotes collaborative working and adopts the ISO 44001 international standard for collaboration. As such it offers an alternative to the more traditional fragmented and aggressive culture born out of the fear of disputes and desire to increase returns through extras. FAC-1 can also provide a consistent pipeline of work that is desirable for manufacturing-led approaches to delivery.

construction frameworks, but the review found that not all of those clearly identified a realistic pipeline of work, nor did they always set targets on how improvements in performance should be measured over time. The review also noted that there is on average only a 20–25% chance of a supplier winning a framework project, and that the average bid cost for a major framework was nearly £250 000 for contractors, going up to £1 m in some cases. That means that if only 1 in 4 bids are successful, a contractor needs to recover between £1 and 4 m before they receive any commercial benefit from a framework. As such, the review put forward a number of recommendations on how the cost of bidding could be significantly reduced by providing consistency for both clients and bidders through the use of standard clauses and forms of contract. Not only does the use of standard clauses reduce the potential cost of bidding as confidence in the responses provided increases, it is also likely to reduce risk and the offsetting of risk to others who may be less well positioned to minimise it.

The *Construction Playbook* (HMG, 2020) identifies a number of tools and techniques to deliver whole life value through construction projects and programmes and requires public projects to follow these on a 'comply or explain' basis. One such tool is the early definition of required outcomes in the form of a project/programme outcome profile (POP), which should be referred to within contracts and used throughout projects to ensure that a focus remains on what the client is really looking for. This is used to identify project and programme outcomes, outputs and metrics, and should be kept up to date as a project develops. While this is to be used for public projects, and typically projects of a reasonable size and complexity, the principles of clearly and succinctly defining required outcomes is a useful practice for projects of any size. It is also important to highlight that the Playbook rightly identifies the importance of passing its principles throughout the supply chain, something that can happen regardless of project size and scope.

While Section 18.5 covers the costing of projects in more detail, it is important to emphasise that, with a move towards considering the whole life value of an intervention, delivery cost is less likely to be the main criteria by which a bid is won or lost. Even where cost has a low weighting compared with other factors used for deciding a bid, cost is still likely to be a deciding factor. The reason is how price is evaluated. If a tender is carried out without any opportunity for post-award negotiation, there is a good chance that the winning tender price is not what the client will end up paying for the work. Of course there are a number of reasons for this, but another key factor is that bids are likely to rely on a number of assumptions. Post-award negotiation is much more likely to close the gap between estimated and final costs.

The Playbook requires any client organisation adopting it to refer any bids that are more than 10% lower than the average of all bids or a should cost model (SCM) to ensure that it has met all the defined requirements. An SCM is a pretender estimate of what an intervention should cost over the whole life of its product (for example, the delivery and operation of a built asset through its design life). This gives the client team a yardstick to measure any potential bids against, with the client's knowledge of what is fully required to meet their objectives. This should be accompanied by benchmarks from previous or similar projects on performance, programme and any other key targets. As with the POP, the principles of benchmarking and SCMs are useful for interventions of any scale and complexity and provide a sense check that what is being proposed and tendered is realistic, achievable and fair. They also provide the client with confidence that the winning bid team know what they are doing.

As with other principles identified in the Playbook, the idea of procuring on best value should be taken through the supply chain. In procurement language, the term most economically advantageous tender (MEAT) is used to describe the offer that provides the best overall solution when considering all criteria (not just cost), and that principle can be used to procure any product or service within the delivery of an intervention. It is a principle used in outcome-based procurement, which is described in Box 18.2.

Box 18.2 Outcome-based procurement

Outcome-based procurement involves clearly defining required outcomes and then procuring a provider based on their capability to deliver against those outcomes, rather than defining how an outcome should be achieved. It can be used for large, complex or, equally, smaller procurements. For example, in construction it can be used to procure building maintenance where a trade may be better placed to define what is required to repair an element than the client. In 2019 the Industrial Strategy Challenge Fund (ISCF) project 'Digitally Connected Supply Chains' (CPA, 2019) used an outcome-based procurement approach to enable potential bidders to offer solutions at multiple levels of complexity and across traditional work packages. The solution had five main steps

- *define* – clearly describing requirements and relationships with other requirements
- *procure* – managing requirements (including BIM data), managing offers and validating requirements
- *deliver* – delivery of the product or service and associated data
- *verify* – receipt and acceptance of delivered products, services and data
- *pay* – automatic payment in line with agreed terms once product or service have been verified.

This approach can be completely open to any organisation to bid against defined requirements, or can be by invitation only, depending on the client and circumstances.

Once requirements have been clearly defined and solutions are developing, the Playbook asks for a delivery model assessment (DMA), which is an analytical, evidence-based approach to assessing how a client should structure the delivery of a project or programme; in other words, once it becomes clear what is to be delivered, how it is best delivered needs to be considered. The key steps in carrying out a DMA are

1. frame the challenge
2. identify data inputs and potential delivery model approaches
3. consider your strategic and operational approach
4. assess the whole life cost of the project
5. align the analysis, reach a recommendation
6. design an effective commercial strategy.

Figure 18.1 describes the key delivery models that the Playbook identifies. It is important to note that these are solution-agnostic, and so it is not the case that only the product mindset approach can take advantage of manufacturing-led construction. Nevertheless, the models described are worthy of reflection for any project.

Figure 18.1 Key delivery models identified in the *Construction Playbook* (HMG, 2020)

Transactional
Traditional approach in which potential suppliers provide a service, and are selected through competition

Hands-on leadership
Where projects are complex, and the client needs greater control

Product mindset
A focus on repeatability and DFMA. Likely to depend on a visible pipeline of work across several projects or programmes

Hands-off design
Where the client is clear on the desired outcomes but agnostic on how it is achieved

Trusted helper
Where the supply chain knows more aware of what needs to be delivered than the client, and a close relationship between the two is required

18.3. Procuring for manufacturing-led construction

The *Construction Playbook* (HMG 2020) promotes the use of manufacturing-led construction in the form of product platforms. While this is to be commended and platforms can deliver significant value, as described in Chapters 13 and 14, there are other approaches to manufacturing-led construction than product platforms and they can require a significant pipeline of work to make the investment worthwhile. The FAC-1 framework alliance contract can help in providing a consistent pipeline for manufacturing-led solutions of all kinds so offers flexibility of solutions and can include suppliers as part of the alliance.

The use of modern methods of construction (MMC) is a core policy within the NHS, where the NHS MMC Toolkit is used for the Procure 23 framework (ISG, 2023). The toolkit provides a well-rounded approach to the procurement of manufacturing-led construction and starts by identifying key required benefits and constraints through the use of scorecards. It then requires detailed responses on the use of manufacturing-led and other technologies as percentages of total construction and their relevant costs. The toolkit provides a simple, consistent and realistic perspective on the level of manufacturing-led technologies used on a project and is preferable to the use of premanufactured value (PMV) as an assessment of such technologies alone.

For smaller interventions, product platforms and framework contracts are unlikely to be used, but that does not mean that manufacturing-led solutions are not suitable or cannot be procured for those. As described in earlier chapters, manufacturing-led solutions do not always need to rely on a steady pipeline of repeat work from Government or another major client to be suitable or beneficial and should not be restricted to use on large projects alone. They do, however, benefit from an outcome-based approach to procurement as it opens the door for new technologies to be used to fulfil a requirement that may not otherwise be considered. Manufacturing-led solutions can be used in any of the delivery models identified in Figure 18.1, and with any common form of contract, but they undoubtedly deliver most value when the commercial model and delivery model are developed alongside each other with a focus on delivering whole life value to the client.

18.4. Automating procurement

The *Construction Playbook* identifies the need for suppliers (in this instance used in the broadest sense to include contractors, subcontractors and consultants) to invest in automated, digital processes including digital payment and contracting systems. This promises to improve transparency, information exchange, payment performance and contract management, as can be witnessed in other sectors that are already more digitalised than construction.

For the purposes of discussing automation of procurement, this section is structured around the five steps identified in Box 18.2, with the addition of onboarding of suppliers.

18.4.1 Onboarding

Onboarding of suppliers is a crucial step in managing any supply chain and should include a number of checks, such as for security and financial stability. However, these checks can be time-consuming if carried out properly and so can often be overlooked, especially for the procurement of small products or services. The accumulative impact can be significant; known as the long tail, when all procurements across a project are looked at together, analysis has identified that for projects valued in the region of £1.5 m, 10% of the project value can be procured through items of less than £5000 each, which is 59% of the procurements carried out (BIS, 2013). For projects over £25 m, this can be 1% by value and 25% by number of procurements. Automation technologies can be used to provide checks for procurements of any size and so are particularly useful for the long tail where resources may not otherwise be spent on checking or comparing offers. Crawler technology can be used to scout supplier websites for key information and for wider searches on previous performance or commentary. Performance on previous projects for the same client or supply chain can be analysed for consideration using analytics, and offers easily compared with bids by other suppliers or on previous projects. As a result, the decision as to which supplier to go with can then become better informed and, at the same time, the consideration of past performance for new work can act as an added incentive to perform well.

Natural language processing (NLP) can be used to read through nondisclosure agreements (NDAs), and robotic process automation (RPA) can create new NDAs. Finally, chatbots can be used for simple supplier engagement and response activities, answering straightforward questions from either supplier or client, as they are commonly used in other industries such as consumer products.

18.4.2 Define

As discussed in Section 18.5, benchmarking and should cost modelling (SCM) are important tools in understanding what performance can be expected and what that performance is likely to cost, with a focus on whole life value. Artificial intelligence is ideal for analysing large volumes of data from previous interventions and recognising patterns such as spend behaviour, and making predictions based on those findings. AI is particularly useful as it can learn the similarities and differences between interventions instead of simply comparing projects from the same sector as though they were like for like. nPlan, described in Chapter 19, provide an example of this, where a significant volume of previous delivery programmes is assessed to predict risks in proposed project programmes.

Previous chapters have discussed requirements management and definition, and, especially at an asset level, accumulative requests can become significant. NLP can be used to analyse both requirements documentation and contract documents to quickly and automatically summarise key requirements or issues. While human input is still recommended, this can significantly speed up the process.

18.4.3 Procure

Chatbots can be used to support procurement, particularly of small procurement items. They can be used for guided buying to help procurement teams quickly identify the right product or service but can also be used to answer common product or service questions. RPA can also be used to create purchase orders and purchase agreements and to begin processing orders of standard goods and services ready for delivery.

In recent years there has been a growing interest in the potential use of blockchain technology in procurement as it can enable transparency and visibility of all transactions across projects and supply chains. It is a technology that makes a decentralised record of transactions in a network that can be managed by third-party network participants, and records of transactions are shared across the network, leaving a public trail of information that makes tampering impossible. While blockchain does have many potential benefits in procurement, there are other solutions that can achieve the same results in a more efficient manner; it is significantly more expensive than solutions such as RPA or supply chain digital twins (SCDTs), which can also track and record all transactions, and scalability can be challenging because transactions are processed more slowly than regular database solutions because many computers need to confirm each transaction. One benefit of blockchain is its independence from the parties to a transaction, but this can also be achieved through the project-wide application of suitable procurement platforms.

18.4.4   Deliver

An SCDT model can be used to record the real-time condition, location and state of product and service transactions. Such a model can also be used to identify and track changes in requirements for smaller transactions, and service agreements can then be automatically updated. Such models are used in the health care and transportation sectors and, by integrating these procurement data, up-to-date cost and programme forecasts can be delivered.

18.4.5   Verify

Irrespective of the form of procurement used, it is clearly important to be able to confirm that a product or service has been delivered and accepted, as it is within any industry. The checking of delivery notes against what has been procured can be semi-automated, although it is still likely to require human involvement to check quality and that what products were due to be delivered have actually been delivered. Most procurement platforms in construction do not extend beyond the point of purchase, so verification of delivery is not a part of the service. In a supply chain digital twin model, however, as utilised in the project described in Box 18.2, payment is linked to the delivery of a product or service. That means that delivery needs to be electronically accepted and verified before the payment process is automatically started (in line with whatever payment terms have been agreed).

Chapter 19 discusses means of automating the tracking of progress of projects on site, such as the use of computer vision to analyse photographs or scans.

18.4.6   Pay

Bots can be used to automate the preparation of invoices and receipts and to check invoices against contracts and delivered products or services.

While this section on automation of procurement has been split into several sections, ideally the whole process will be pulled together into one platform. A good example of this within the construction industry is ProcurePro, described in Box 18.3.

Box 18.3 ProcurePro

ProcurePro is a procurement platform specifically for the construction industry, aimed at organisations appointing subcontractors. It can be used to connect up every step of the procurement process, such as scope management, tenders, comparisons, approvals and contracts, and provides real-time visibility across projects and supply chains. It incorporates an extensive library of scope of work items, which enables sharing of lessons learned, visibility of performance across multiple projects and vendor management tools to quickly understand how specific contractors have performed previously and what their capabilities are.

18.5.   Benchmarking and cost modelling

A key aspect of procurement is understanding what a transaction is likely to cost and how whatever is being procured will likely perform, not only in terms of building performance but also in areas such as embodied carbon. Depending on the information available and the size, complexity and uniqueness of the intervention, this can be achieved through benchmarking, SCMs or cost modelling.

18.5.1 Benchmarking

Benchmarking is the process of analysing evidence of cost or performance of similar interventions to predict the likely implications of a proposed project and is usually carried out at the early stages of a project to help support investment decisions. It is critical that it is carried out with caution; no two projects are the same, and using inaccurate or inappropriate data for benchmarks can lead to the wrong decisions being made on whether to take a project forward or not. To this end, it is preferable to have subject matter experts involved in the benchmarking process to help clearly define the similarities and differences between projects and to quantify the impact these differences will make on any comparison. Benchmarking is usually only formally carried out on relatively large projects, and for these purposes the Infrastructure and Projects Authority (IPA, 2021) have developed a process and guidance document to support its application. However, informal benchmarking is carried out on projects of all sizes – for example, estimating the cost of a domestic project based on a loose comparison with a previous project by the same builder, same consultant or in the same area. There is a vast difference between this and a formal benchmarking exercise, however, but the use of artificial intelligence can bring the informal process much closer to the formal one. For example, machine learning can be used to analyse large volumes of data from previous projects by the same organisations and quickly understand the differences and similarities between them. Whether automated or not, the key steps in a benchmarking process identified by the IPA (2021) are relevant, namely

1. confirm the project objectives and set of metrics
2. break the project up into major components for benchmarking
3. develop templates for data gathering
4. scope sources and gather data
5. validate and re-base data
6. produce and test the benchmark figure
7. review and repeat as necessary.

There are fundamentally two approaches to benchmarking. The first is top-down, where the benchmark starts with the total cost of a project, which is then broken down into smaller components. This approach is most suited to early-stage comparisons. The second is bottom-up, and relies on more granular sources, such as the cost of labour and materials. To use this method relies on having a design to assess, however, so is less useful for early-stage benchmarking. However, it may be useful where there are no suitable benchmarks at a project level or there is known to be a spike in labour or material costs.

Tools such as QuickEst by Causeway can be used to create quick and accurate cost benchmarks using data from previous projects.

18.5.2 Should cost models (SCMs)

As mentioned in Section 18.2, SCMs are pretender estimates of what a project should cost over its whole life (delivery phase plus its full design life). They include additional market factors, such as risk and profit for the supply chain, and provide an early, realistic indication of what a client can expect to pay for the delivery and operation of an asset based on delivery by a reliable and competent supply chain. As with a bottom-up benchmark, an SCM requires a design of some sort to be in place, and can be carried out a number of times as that design develops. Assessing the likely cost across the asset's life cycle differentiates an SCM from a traditional cost model and provides a more rounded assessment of potential solutions. As with benchmarking, SCMs are more likely to be carried out on projects of a reasonable size and complexity, but there is no reason why they cannot be used on projects of any size, if the desire is there and the skills are available to carry out such an assessment. While tools exist to automate much of the delivery phase aspects of an SCM, assessing the through-life cost is a harder prospect. However, it is anticipated that machine learning will be useful for this in the near future.

18.5.3   Cost modelling

Cost modelling is the process of providing costs for a project based on a proposed solution and can cover different project phases of an asset or can incorporate the whole life of an asset. For the purposes of this section, the focus is the delivery phase.

Cost modelling is likely to vary in levels of detail and accuracy at different stages. However, it does rely on a design from which quantities and some form of specification can be taken. Cost modelling is usually carried out by a specialist consultant, but the use of digital tools can significantly speed up and support their activity. For example, the viability tool mentioned in Section 16.2 uses quantities and a predefined specification to provide an accurate cost model based on project inputs and assumptions. KOPE also develops a cost for the product categories applied based on accurate quantities taken from design models.

While if an accurate BIM model exists, schedules of quantities can readily be produced, there can still be a separation between extracting the quantities and producing a cost model. However, tools such as Cost-OS exist (described in Box 18.4), where the BIM and cost model can be directly linked. Not only can quantities be taken directly from the BIM, but also 2D plans can be combined to create 3D models and quantities.

Box 18.4 Cost-OS

Cost-OS by Nomitech is an intelligent cost modelling platform. It enables quantities to be taken automatically from 3D models in minutes, enabling accurate and updatable cost models. Where 3D models do not exist, 2D files can be used for semi-automated on-screen take-offs, which can also produce fully classified 3D models. Cost-OS includes an intelligent Assemblies Development Module, which makes it very easy to quickly and reliably create a detailed, parametric top-down model from initial designs. Alternatively, a bottom-up model can be created from base rate build-ups. Cost-OS enables multiple users to work on the same estimate in real time, and fully resourced and costed models can be exported in IFC (industry foundation classes).

Tools such as Cost-OS and others are very useful where BIM or even accurate 2D files are available, and the level of automation will depend on the accuracy and format of those files. However, for smaller projects where either there are no drawings or what does exist is in hard copy format, it can understandably be more challenging, time consuming and error prone to extract quantities. Tools such as Sort It AI (described in Box 18.5) enable users to scan or take a photograph of scaled drawings, and produces a schedule of quantities of elements. From this, or from user voice or text input, the tool then enables prices and stock levels of lists of products to be obtained from, and compared between, a number of local builder's merchants. With over 99% of businesses in construction being small businesses, tools such as Sort It AI which aid in the digitalisation of requirements and procurement directly address parts of the market often overlooked when it comes to the adoption of automation.

Box 18.5 Sort It AI

Sort It AI is a quantities take-off and price comparison tool aimed at small builders and subcontractors. There are two main capabilities within the platform, the first of which is the ability to take quantities from images of plans. Using computer vision, quantities can be taken from scanned images or photoraphs of drawings, and a list of components and areas can be generated. This can then be refined to create a list of products and materials required.

The second aspect of the tool is price and stock comparison. A list of required products can be created or edited from the list generated from the images, and within minutes price and stock comparisons from local builder's merchants can be generated. This enables accurate, up-to-date estimates to be delivered within minutes, all from a phone. Figure 18.2 illustrates different steps within the app, including comparing drawings, building estimates and prices.

Figure 18.2 Sort It AI takes images of plans, calculates and creates a bill of quantities, compares prices across local merchants and enables orders to be placed (Source: Sort It AI)

18.6. Supply chain management

The *Construction Playbook* emphasises the need for public spending to flow down through the supply chain and to support the growth and inclusion of SMEs. To enable that to happen, lower tiers of the supply chain need to be visible and to have the opportunity to influence their scope of work. As mentioned in Section 18.1, ESI can be beneficial to both clients and supply chain partners as it provides the opportunity for those who are likely to deliver a scope of work to share their experience and knowledge and, as a result, to reduce unnecessary costs and risk. It can be difficult for SMEs to have access to projects at their early stages, however, but with the use of agreements such as FAC-1 they can either become involved at a framework or an individual project level if they have desirable skills that are of value. Another approach can be to use an outcome-based procurement model such as that described in Box 18.2, where any preapproved supplier can bid for any tender that matches their capabilities.

Managing the supply of products and services on large, complex projects can be a significant challenge, and one that digital solutions and automation can certainly assist with. At Hinkley Point C, the Wincanton Supply Chain Integrator (WSCI) is used, which is described in Box 18.6.

Box 18.6 Wincanton Supply Chain Integrator (WSCI) by Wincanton

Wincanton are a logistics and supply chain management business operating across several industries, including construction. They provide logistical services to a number of building product manufacturers and businesses. Their WSCI capability is an example of a logistics and supply chain management platform which aims to make the complex construction supply chain visible and transparent through tracking and reporting. All shipments can be tracked from purchase order through to delivery on site, and all material supplies can be managed on one system. A partner portal is used to manage supplies, delivery slots and storage and to manage documentation, communications and reporting. WSCI has been used on the development of the Hinkley Point C power station.

SCDTs also provide opportunities for the management of construction supply chains. Although they are not widely adopted in the construction industry yet, it is anticipated their use will grow between now and 2035. SCDTs are detailed models of actual supply chains, using real-time data or snapshots of the state and location of supply chain products and services. They integrate live information feeds, shipment and payment schedules and inventory levels, with IoT and GPS feeds, and can enable the segregation or aggregation of procurements to be visualised and analysed on the fly.

The construction industry is largely built on a model that maximises flexibility of resources, where contractors of all types can grow and contract in line with demand. It is important to recognise that when moving towards a more manufacturing-led approach to construction the impact on the whole supply chain needs to be considered. If only a part of the supply chain is aggregated and the rest remains fragmented, many of the same issues that the industry currently faces can still arise; it is just that the position of those challenges may change. But also, when integrating solutions it is likely that some of the flexibility that the traditional model benefits from can be lost, unless the integrated solutions also have the ability to grow and contract in line with demand. This is where the development of an integrated solution takes great care and thought, and where some volumetric businesses have failed in recent years.

18.7. Technology relevance summary

Using the modelling framework described in Chapter 6, Figure 18.3 summarises the potential relevance of different automation technologies to supply chain management and procurement activities. For this purpose it is assumed that the majority of procurement occurs in the Technical Design and Manufacturing and Construction stages of a project. However, the Plan and Source stages also incorporate the use of frameworks.

Figure 18.3 Technologies relevant to supply chain management and procurement (Author's own)

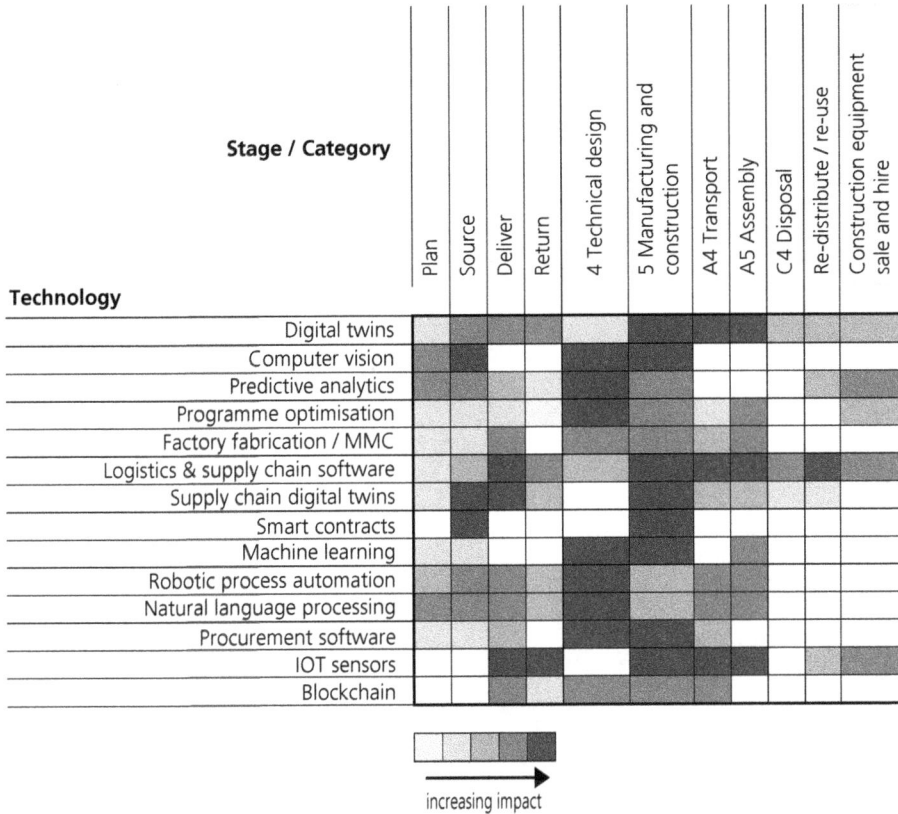

Stage / Category columns: Plan, Source, Deliver, Return, 4 Technical design, 5 Manufacturing and construction, A4 Transport, A5 Assembly, C4 Disposal, Re-distribute / re-use, Construction equipment sale and hire

Technology rows:
- Digital twins
- Computer vision
- Predictive analytics
- Programme optimisation
- Factory fabrication / MMC
- Logistics & supply chain software
- Supply chain digital twins
- Smart contracts
- Machine learning
- Robotic process automation
- Natural language processing
- Procurement software
- IOT sensors
- Blockchain

→ increasing impact

REFERENCES

Alliance Forms (2024) https://allianceforms.co.uk/fac-1/ (accessed 2/8/2024).

BIS (Department for Business Innovation and Skills) (2013) *Supply Chain Analysis Into the Construction Industry*. BIS, London, UK.

Constructing Excellence (2011) *The Business Case for Lowest Price Tendering?* Constructing Excellence, Watford, UK.

CPA (Construction Products Association) (2019) https://www.constructionproducts.org.uk/media/537687/supply-chain-digital-twin-release.pdf (accessed 2/8/2024).

HMG (Her Majestey's Government) (2020) *The Construction Playbook*. The Stationery Office, London, UK.

IPA (Infrastructure and Projects Authority) (2021) *Best Practice in Benchmarking: Government Project Delivery Framework*. IPA, London, UK.

ISG (2023) New NHS toolkit for MMC to drive cultural change across UK healthcare. https://www.isgltd.com/uk/en/news/new-nhs-toolkit-for-mmc-to-drive-cultural-change-across-uk-healthcare (accessed 01/08/2024).

Mosey D (2021) *Constructing the Gold Standard: An Independent Review of Public Sector Construction Frameworks*. Cabinet Office, London, UK.

Steve Thompson
ISBN 978-1-83608-599-7
https://doi.org/10.1108/978-1-83608-598-020241022

Chapter 19
Integration automation

19.1. Introduction

Within the whole life cycle of an asset, the delivery stage is potentially the least automated and most traditional of all stages. With the rapid development of technologies such as ChatGPT and the use of chatbots when looking to purchase something online, in many areas of modern life it is reasonably easy to see that automation can have a significant impact, but in the physical building of something it can be harder to imagine. However, the delivery phase involves much more than the physical delivery alone; there is the observation and analysis of the site and environment, the detailed planning and management of tasks, the preparation for construction activities, the construction itself and the reporting of progress, all of which offer opportunities for automation. Automation has the potential to significantly improve productivity, improve collaborative activity planning and logistics management, especially when combined with manufacturing-led construction. The use of connected automated plant (CAP) alone can potentially lead to productivity improvements exceeding £200 bn by 2040 and a 37% reduction in construction fatalities (TRL, 2020). The same technology can potentially produce cost savings of over £125 m on a 50 km dual carriageway road construction, and the use of telemetry data has the potential to save over £25 m annually through reduced utility strikes (TRL, 2020). While the application of such technologies is growing (for example, all National Highways sites must now use 3D machine control (3DMC) for all earthworks operations), there are still potential barriers to their wider use. These include technological challenges (which will be discussed in this chapter), potential costs and perceived risks, and lack of specific regulations, but these barriers are expected to reduce over the medium term (TRL, 2020).

19.2. Observing the environment

Measuring and locating elements in the local environment are key activities when starting the delivery phase of an intervention, and ones that can readily be supported by technology. While the scanning of work as it progresses is covered in Section 19.9, and the measurement and modelling of existing assets is described in Chapter 20, technologies can also enable the visualisation of features that cannot be seen with the naked eye. For example, Emerald Geomodelling carry out geo scans of large sites by helicopter, which penetrates the ground by up to 100 m to reveal ground conditions and bedrock topography. These surveys can be carried out at a rate of 90 km per day, and are followed up by strategically placed drilling to develop accurate 3D models of ground conditions. Another promising technology development is the National Underground Asset Register (NUAR). This is a digital map of underground pipes and cables across the UK. NUAR is currently under development and is expected to be operational in 2025. It promises to deliver economic growth of £490 m per year through increased efficiency, reduced disruptions to the public and reduced asset strikes (DSIT, 2022).

Figure 19.1 Scanning technologies to automate the measurement of existing spaces for the accurate production and fitting of a kitchen worktop (Author's own)

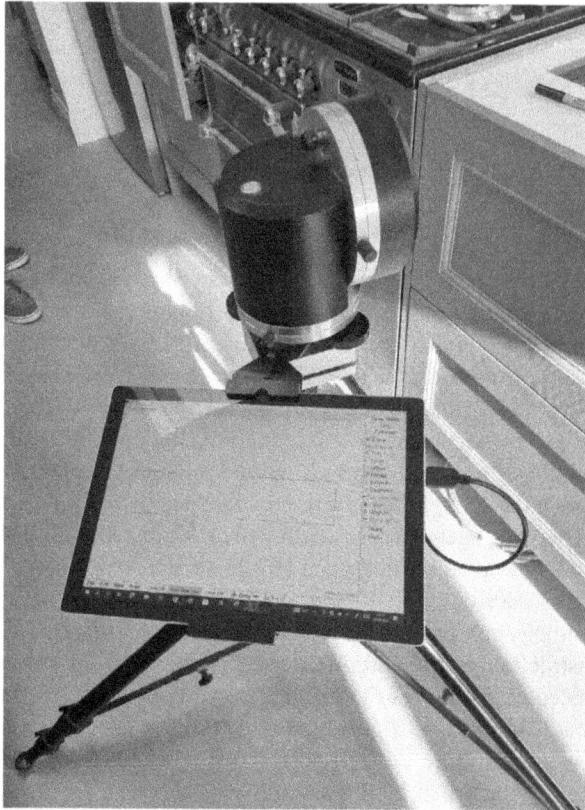

Observing the environment is not only a requirement for large projects. At the other end of the scale, most kitchen worktops today are measured using lasers to speed up the templating process and to provide accurate results that can be exported to cutting machinery. Figure 19.1 shows such a templating process being carried out.

19.3. Programme automation and 4D modelling

Construction programmes are crucial tools in the delivery of anything but the simplest of interventions in that they help structure the delivery of a project and coordinate trades and activities, identify potential risks to programme and provide forecasts on programme, resource requirements and potentially cash flow. On complex projects, plans can run into hundreds or even thousands of lines of activities and become complex and unwieldy, and it can be difficult to spot potential risks ahead of time. However, there are tools that can help automate aspects of planning, given the correct inputs.

The first aspect that can benefit from automation is the creation of a baseline programme. Tools such as ALICE Pro (described in Box 19.2) enable users to create templates for different construction types and apply those to BIM models to generate detailed construction schedules for consideration and analysis. However, not all projects have a BIM model, and so for smaller projects there are fewer options to automate the generation of programmes, but there are possibilities. The basic principle is the same as in the ALICE solution, which is to be able to combine a construction logic template (which describes the details of how something is constructed, but without specific quantities) with a project-specific form of quantities (for example, an elemental breakdown of a project or a BIM). The viability tool developed by PCSG and Wienerberger described in Chapter 16 was based on this logic, and it is how many projects are traditionally estimated – a combination of experience and quantities. It is always likely that these forms of automation will require some revision or editing to be suitable for construction stage schedules, however, in order to deal with any project specific requirements that are beyond the scope of a template. Beyond this form of simple programme creation, spreadsheet tools such as Microsoft Excel can be used to create project schedules for smaller projects, but this is much more of a manual process of inputting lists of activities, durations and resources. There are many templates available to help with the process of creating schedules in this way and, while certainly nowhere near as thorough or reliable as professional planning software and solutions, they are more flexible and informative than manually creating a static programme from scratch that does not have relationships between activities incorporated.

The second aspect of programme development that can be automated is analysis, optioneering and monitoring. Box 19.1 describes Insights Pro by nPlan, which is a machine learning tool that uses data from past projects to assess the risk of schedules inputted into the application, and offers suggestions for optimisation and mitigation of risk. Such tools are incredibly powerful on large, complex projects in particular and can save time, money and resources based on detailed learning from a vast resource of past deliverables. They can quickly analyse large volumes of activities and identify issues that may otherwise be overlooked as being relatively unimportant.

Box 19.1 Insights Pro by nPlan

Insights Pro is an AI-driven forecasting and risk management tool that has been trained on over 750 000 past construction schedules to identify potential risks and opportunities to mitigate them. It clearly identifies critical paths and carries out schedule integrity analysis to find potential areas for improvement. The tool identifies the likelihood that specific tasks will be completed on time and the potential time savings through mitigation of risk for specific tasks. Insights Pro also has a large language model named Barry, which acts as an AI assistant that can answer questions on your schedule once it has been analysed, such as suggestions on how to mitigate risks and producing project summaries.

Four-dimensional modelling (or 4D planning) is the use of 3D modelling linked to construction schedules to enrich the construction planning process by visualising results. They can be high-level models at the early stage of a project, through to very detailed and comprehensive models with animations of activities at a detailed design or construction stage. A 3D model on its own does not take into account the process of construction or the intermediary steps involved in the construction of an asset, which is where a 4D model is of use.

Box 19.2 ALICE

ALICE is a planning optimisation engine which enables users to optimise construction schedules and compare different scenarios. There are two different methods that can be used; with ALICE Core, users upload their existing schedules, and the tool then simulates and optimises them by recommending potential improvement. Different variables can be edited, such as adding more of a particular resource, and the results compared with the baseline schedule. With ALICE Pro, users upload a 3D BIM model instead, and the tool automatically generates a baseline schedule based on the construction methods and resources that you have identified, ready for modification and analysis. It can then be visualised and edited in a 4D modelling environment.

It is best to start 4D modelling as early as is practical on complex projects in particular in order to aid in the planning of construction activities and the buildability of projects. As earlier chapters have highlighted, the majority of construction work occurs at the lower tiers of the supply chain, and 4D modelling (especially when combined with interactive mission rooms) can provide opportunities to engage with those involved in delivery and benefit from their expertise. Tools such as ALICE, described in Box 19.2, enable project schedules to be generated from 3D models and construction templates, and then different options to be compared and optimised. While many tools automating construction programming require specialist skills and software, they can still be used on projects of any size if the right capability exists.

19.4. Connected autonomous plant (CAP)

Connected autonomous plant (CAP) (or connected automated plant as it is also known) is construction equipment that can partially or fully operate automatically with reduced human input. While still not widely used on construction sites, it is gaining in popularity and is expected to significantly increase over the next decade. Semi-autonomous plant is used extensively in mining and ports industries, where systems typically follow a predetermined route with little capacity for deviations. In construction, CAP needs to be more flexible, and may not be able to rely on predefined routes. For example, CAP can be used to excavate or cut to grade where it combines accurate location information with digital terrain models. In such uses, a defined route is not necessary; what is important is the end result. In 2020 National Highways (then Highways England) commissioned a review and road map aimed at making the use of CAP the norm by 2035 (TRL, 2020). In addition, PAS 1892 (BSI, 2023) was published in 2023, providing guidance on how to define and specify CAP. Both documents provide a framework for identifying the level of automation of CAP, which provides a more granular assessment than a simple single value, as the level of automation is assessed based on five key areas. This is important as it enables plant of different types to be readily compared. The five areas are

- *observe* – the act of gathering information on the current state from the surrounding environment through various sensors and communication channels
- *understand* – processing the data from observation, including comparing, learning and predicting
- *decide* – developing a set of possible actions that could be carried out, then deciding which action to take

- *act* – carry out the selected action
- *responsibility and fallback* – this is the process of defining who or what is responsible for ensuring that when plant suffers a failure, appropriate action can be taken. Includes monitoring.

These categories also provide useful prompts when analysing the wider level of automation and appropriateness of construction robotics to a given task.

Machine control is the most widely used CAP system to date, and is a collection of systems that know the position of plant and its functional components (for example, an excavator, the position of its arm and bucket). Box 19.3 describes the RPD 35 by Built Robotics, a CAP which uses machine control to position and drive piles.

Box 19.3 RPD 35 by Built Robotics

The RPD 35 is an autonomous pile driving system primarily for solar farm developments. The system utilises a standard excavator, with baskets to carry up to 200 piles. It also includes a picker and hammer, and the brain of the system. Using GPS, the robot can accurately install over 300 piles per day, compared with approximately 100 using traditional means. The pile plan is uploaded to the robot's computer, and automatically divided into sequences based on the number of piles it can carry. It then installs all the piles in its load, and returns to the starting point to be loaded up ready for the next sequence.

Built Robotics also produce the Exosystem, which automated the operation of diggers for trenches.

There are significant differences between automating plant and machinery in a controlled industrial environment and on a construction site. For one, in an industrial environment it is common for robots and machinery to remain static, and for materials and humans to be brought to them to carry out tasks. On a construction project it is more likely that the machinery needs to move around to carry out tasks. However, that does not mean that the plant or robot should be allowed to move without any control or limitations. As such, PAS 1883 (BSI, 2020) defines a specification for an operational design domain (ODD). ODDs provide restrictions on where, when and under what conditions a CAP is designed to operate in. Typically this will include location boundaries, and if the CAP moves outside of this or any other specified boundaries, it can be reported or stopped accordingly.

To enable machinery or robots to operate autonomously, both they and humans need to know exactly where they are on a site. Technologies such as GPS and cellular networks alone are not sufficient as they do not operate in three dimensions (so height within a development cannot be tracked), and they are also subject to interference or blackouts within structures. For these purposes, real-time location systems (RTLS) technologies such as Quantum RTLS can be used (described in Box 19.4). In addition to providing incredibly accurate locations, they can display results in real time, which can be crucial to support effective decision making.

Box 19.4 Quantum RTLS by ZeroKey

Quantum RTLS is an example of a real-time location system (RTLS) and is accurate to 1.5 mm in three dimensions. Many location systems can provide locations in two dimensions, but for assets of more than a single storey, they are not suitable. The system can be used to track machinery, people or products across a construction site and does not rely on broadband or cellular connectivity. It is based on acoustic signalling, similar to how bats use echolocation to navigate. The Spatial Intelligence Platform that works alongside the location system displays precise locations in real time, and can trigger events such as warnings of potential collisions.

19.5. Robotics

As described in Section 19.4, the suitability and use of robots on construction sites is very different from industrialised, controlled environments. Not only is this due to the fact that machines are likely to move in some way, but also the number of tasks that are likely to be undertaken simultaneously by different parties can limit their effectiveness. For example, if a robot is used to carry out a repetitive task, but requires an area of the site to be otherwise vacant, it can have a negative effect on the overall productivity of a project as other tasks need to be put on hold until the robot is finished. However, where repetitive tasks are required and they are carefully considered as part of a delivery plan, there is no question that on-site robotics can help improve construction productivity and safety.

There are two main types of robots used in construction. These are

- *stationary robots*
    - *gantry robots* – an overhead system with a connected manipulator, commonly used to pick and place products
    - *robotic arms* – otherwise known as articulated robots, they are typically used for tasks such as welding, handling, painting, drilling and waste separation
    - *cable robots* – these can cover large areas and work on a similar system to overhead spy cameras used at sporting events for overhead shots
- *mobile robots*. These can create larger structures than stationary robots as they are not restricted by their size, including
    - wheeled robots
    - guided robots – where guiderails of some form are required
    - walking robots
    - flying robots.

The majority of construction robots are single-task construction robots (STCR) in that they are designed to carry out single tasks, which are usually trade-specific. For example, they can be used for painting, brick laying, concreting or concrete finishing, façade installation, additive manufacturing, welding and maintenance, among other tasks. In total, Bock and Linner (2016) identified 28 categories of STCRs. Box 19.5 describes Hadrian X, a brick laying robot developed by FBR. This is more flexible than many brick laying robots as it does not require a track for the robot to move along, and it can be used for both internal and external walls.

Box 19.5 Hadrian X by FBR

Hadrian X is a brick/block laying robot which averages the laying of 300 cement or clay blocks per hour, or over 1500 standard brick equivalents per hour, compared with the traditional human laying of around 500 bricks per day. Unlike some other brick laying robots, the system utilises a large telescoping boom arm from a stationary vehicle, and so does not rely on a track or other infrastructure being in place before it can carry out its task. The system uses a 3D model as its input, which is created using proprietary software. Two humans work with the machine to load packs of bricks using a telehandler, install brick ties when needed and to monitor for quality control purposes. Figure 19.2 illustrates the Hadrian X and its reach.

Figure 19.2 Hadrian X by FBR is a brick laying robot that can lay bricks and blocks significantly quicker than using traditional methods, can produce internal or external walls, and is not reliant on tracks for the robot to follow (Image © FBR Ltd)

Bricks were developed to be easily lifted and installed by hand, and not too small that half of the wall were made up of mortar. In other words, they were designed around the user and performance requirements. If the same approach were to be used when building with robots, the answer would likely be something larger, heavier and more efficient, which is why advanced brick laying robots such as the Hadrian X can also install larger, heavier blocks than are used traditionally, and why there are now a number of solutions to create the impression of a brick wall while installing large prefabricated panels. However, some alternative brick laying robots without the same capabilities (for example, those requiring tracks, only able to install traditional bricks and limited to external rectangular walls only) are likely to have a limited shelf life. There is a debate as to whether the aesthetics should stay the same or not as new construction approaches are utilised, but to achieve improvements in efficiency and productivity new methods are going to be required. Thomas Bock (Bock, 1989) developed an approach called robot oriented design (ROD) which aims to reduce the complexity of assembly processes by reducing parts count to deliver the desired outcome in a more efficient way using automation and robotics.

While STCRs are focused on single tasks, they do in some ways provide flexibility in that they can in theory work alongside traditional processes and do not require the whole site to be structured and automated. They can be used as part of a semi-automated process in on-site production facilities, such as those described in Section 19.6.

Robots can also be used for temporary works. A good example of this is in the use of climbing robots to help in the construction of high-rise building cores without the use of cranes. SKE plus by Doka is described in Box 19.6, which enables construction working platforms to climb a building's core as it is built.

Box 19.6 SKE plus by Doka automatic climbing formworks

The SKE plus is a formwork system that automatically climbs structures (such as building cores) without the need for cranes. The system encloses the work area to provide a safe working environment at any height, and so is particularly useful for high-rise developments. The system is hydraulic and works as a kit that can be adapted to any shape of layout and structure height. Figure 19.3 illustrates the SKE plus in action.

Figure 19.3 SKE plus by Doka is a formwork system that automatically climbs structures without the need for cranes (Source: Doka)

Climbing robots are different from sky factories, which are described in Section 19.6. However, climbing technology can also be used in sky factories. There are also robots such as FieldPrinter by Dusty Robotics, which is used to print setting out lines on concrete slabs for fit-out work, and Spot by Boston Dynamics which is used primarily to capture images and scans of progress. Because Spot has four legs, it is more agile and able to overcome obstacles than most robots, and it can change route on the fly to cope with dynamically changing environments.

19.6. Temporary site-based factories

Temporary site-based factories are different from flying factories, which are temporary production facilities typically located near to site rather than on the site itself. This section looks at temporary production facilities within the site itself, bringing automation and production directly to the project location.

Some forms of site factories have been commonplace on construction sites for several decades, including rolling machines and framing set-ups for light gauge steel framing, and rolling machines producing aluminium roof profiles that run directly from machines to their position on a roof, ready for final installation. Site factories can be more involved than single operations, however, and can potentially provide a temporary structured environment from which STCRs can work together or with other machinery. Factories can be located on the ground or, as described in Box 19.7 and Figure 19.4, they can be used at height and move up as a building develops. These are often known as sky factories.

Box 19.7 Sky factory by Mace at East Village No.8, London

Six-storey rising factories were used for two residential towers at the East Village in Stratford. Each factory was built around the building's core, and included spaces dedicated to materials delivery, façade installation and assembly of prefabricated components. The factories each weighed 510 tonnes and were 35 m wide by 41 m long and 33 m high. They enabled a floor to be constructed in just one week, and have been described as the UK's first rising factories (Mace, 2024).

In on-site factories, the flow of materials and other logistics need to be carefully considered and work alongside the production facilities to deliver just-in-time results. If an on-site factory produces output that then needs to be stored without good reason, any efficiency gained through production can be lost.

19.7. Workforce management and communication

As has been described in earlier chapters, construction is one of the least digitised industries, but at the same time, the majority of workers on site will carry advanced communication devices with them in the form of smart phones. Communication on sites of any size is clearly key to health and safety, but also to management of resources and ensuring that teams know what they are doing without having someone watching over their every move. It is no surprise that platforms and applications have been developed to aid communication between parties on site, and Box 19.8 describes an example from Rhumbix which supports communication and reporting.

> ## Box 19.8 Rhumbix
>
> Rhumbix is a platform for managing a construction workforce. It can manage timekeeping, production tracking, health and safety reporting and daily performance reporting and equipment tracking. Production levels can be documented on a daily or weekly basis, and change orders can quickly be shared with relevant stakeholders.

JobNimbus, described in Box 19.9, goes a step further and enables those in the roofing trade to manage large portions of their activities in one place, including marketing, sales management and workforce management. Both Rhumbix and JobNimbus make it easier to ensure that workers can log their time and activities simply and clearly in order to enable them to demonstrate their work and get paid accordingly. From a management perspective, it means that a clear record can be produced and be easily accessible.

> ## Box 19.9 JobNimbus
>
> JobNimbus is a software platform aimed at the roofing trade. It provides a wide range of services including marketing, sales, production, communication and billing. It enables tasks to be assigned to team members with identified deadlines and progress tracking, and materials to be procured and managed directly through the software. It also allows multiple projects to be viewed simultaneously so that a workforce can be managed across a portfolio.

19.8. Material and performance management

Similar to tracking workforce movements and activities, the management of materials on any construction site can be time consuming. Tools such as QFlow, described in Box 19.10, make the process easier by enabling workers to simply record material movements (whether material deliveries or waste separation and movement) using their mobile devices and upload them to a central record.

> ## Box 19.10 QFlow
>
> QFlow is a platform for managing material movements and waste, and enables all material deliveries to be recorded quickly through mobile applications and photographs. Because they quickly record that materials have been delivered and waste transferred, invoices can be created and the supply chain paid. Bouygues UK used QFlow and reported a 271% efficiency gain in invoice and payment processes as a result (QFlow, 2024). In addition to confirming delivery or transfer of materials, QFlow also checks supplier and material certifications, such as FSC for timber and ISO14001 for suppliers.

Applications such as SR Measure by Stockpile Reports are also useful in automating the measurement and quantification of stockpiles of materials. Using a suitable smart phone, users just point the camera at the pile and walk around it. This then calculates the volume and weight of the material.

Systems also exist to help in the measurement of concrete strength and temperature as it cures, such as ConcreteDNA by Converge. Remote sensors embedded in the concrete report directly to the cloud for monitoring. The system also includes MixAI, a tool that uses predictive analytics to help contractors to select suitable concrete mixes based on their desired outcomes.

19.9. Progress tracking, verification and reporting

Progress tracking is an area where technology has developed rapidly in the last decade, as reality capture technologies have become smaller, cheaper, more accurate and more portable. The suitability of different reality capture technologies for different purposes is described in more detail in Chapter 20, but for the purposes of tracking progress and valuations of work completed, sub-centimetre accuracy is not usually the main concern; it is more important to recognise whether an object is present or not, and to calculate areas or volumes sufficient for valuations.

Box 19.11 describes technologies by OpenSpace, as illustrated in Figure 19.4. Helmet-mounted cameras quickly capture the environment. Once captured, the reality of the site can be compared directly with the BIM model to highlight any clashes or errors. OpenSpace Track uses computer vision and machine learning to compare progress against the schedule and support valuations. Buildots is another platform that uses a similar reality capture approach and enables users to reassign or prioritise future tasks.

Box 19.11 OpenSpace

Figure 19.4 OpenSpace use helmet-mounted cameras to accurately and quickly scan and record construction sites. Software then maps these scans to models and plans and can enable users to overlay BIM models onto reality captures (Source: OpenSpace)

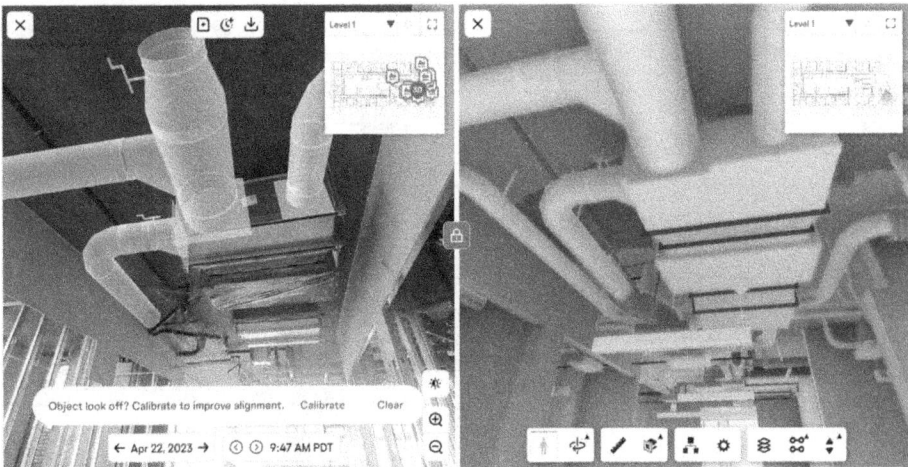

OpenSpace uses helmet-mounted 360° cameras to capture progress as workers walk around a site, whether a quick renovation, an infrastructure or a multi-year megaproject. Lidar-enabled camera phones can also be used with OpenSpace to quickly get a 3D scan with 2-inch measurement accuracy.

OpenSpace is quick and can capture up to 25 000 sqft in 10 minutes, and the vision engine automatically maps images to models and plans. OpenSpace BIM + enables split screen views of reality captures alongside BIM models and allows users to overlay BIM elements onto captured images, making it easy to identify progress and any discrepancies. Field notes can also be exported into the BIM model. OpenSpace can also track progress automatically to quantify work-in-place and compare work completed against schedules.

Figure 19.5 Technologies relevant to integration activities (Author's own)

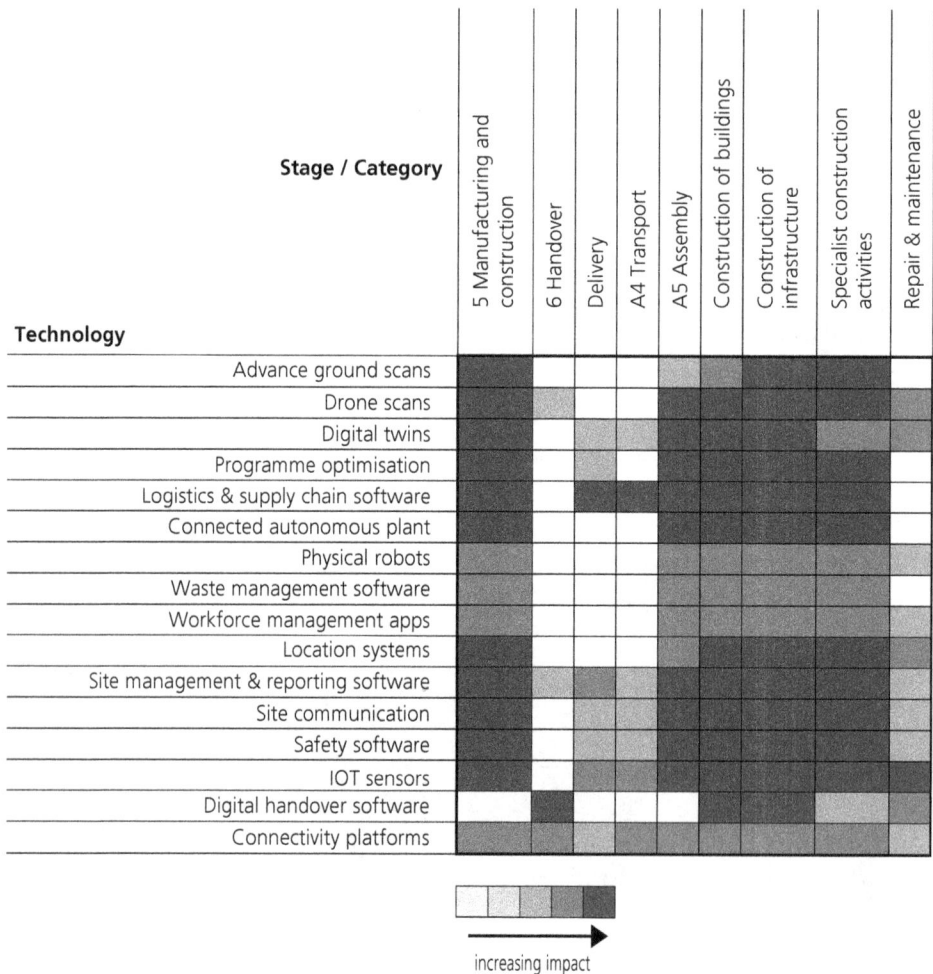

Technology \ Stage / Category	5 Manufacturing and construction	6 Handover	Delivery	A4 Transport	A5 Assembly	Construction of buildings	Construction of infrastructure	Specialist construction activities	Repair & maintenance
Advance ground scans									
Drone scans									
Digital twins									
Programme optimisation									
Logistics & supply chain software									
Connected autonomous plant									
Physical robots									
Waste management software									
Workforce management apps									
Location systems									
Site management & reporting software									
Site communication									
Safety software									
IOT sensors									
Digital handover software									
Connectivity platforms									

increasing impact

For precision capture of an environment to ensure objects are accurately positioned, technologies such as laser scanning or Atom by XYZ Reality are more appropriate. These rely on trained users, but are ideal for setting out and validating that items are exactly where they should be to avoid rework.

19.10. Technology relevance summary

Using the modelling framework described in Chapter 6, Figure 19.5 summarises the potential relevance of different automation technologies to construction delivery activities and management. Technologies relating to understanding and scanning the environment are considered in Chapter 20. When compared with other aspects of asset design, development, delivery and operation, there is potentially a broader range of relevant technologies to this phase. However, adoption is not as high as in other phases, but it is forecast to grow considerably over the next decade.

REFERENCES

Bock T (1989) *Robot Oriented Design.* Faculty of Engineering Dr.-Ing. dissertation, Chair of Building Production, University of Tokyo, Tokyo, Japan.

Bock T and Linner T (2016) *Construction Robots – Elementary Technologies and Single-Task Construction Robots.* Cambridge University Press, New York, NY, USA.

BSI (2020) PAS 1883: 2020: Operational design domain (ODD). Taxonomy for an automated driving system (ADS). Specification. BSI, London, UK.

BSI (2023) PAS 1892: 2023: Connected and automated plant (CAP) – Defining and specifying the use of CAP in sonstruction and maintenance works for the purposes of procurement and deployment – Specification. BSI, London, UK.

DSIT (Department for Science, Innovation and Technology) (2022) National Underground Asset Register (NUAR). https://www.gov.uk/guidance/national-underground-asset-register-nuar (accessed 08/03/2024).

Mace (2024) One Big Leap for Construction. https://www.macegroup.com/projects/east-village-no8 (accessed 10/03/2024).

QFlow (2024) Castle Park View – Bouygues UK. https://www.qualisflow.com/castle-park-view-bouygues-uk/ (accessed 08/03/2024).

TRL (Transport Research Foundation) (2020) *Connected and Autonomous Plant: Roadmap to 2035.* TRL, Wokingham, UK.

Steve Thompson
ISBN 978-1-83608-599-7
https://doi.org/10.1108/978-1-83608-598-020241023

Chapter 20
In-use automation

20.1. Introduction

While the construction phase of an asset's life cycle has a significant impact on its whole life energy use and carbon emissions, the operational phase is still the largest contributor. Of the 39% of total global carbon emissions that come from construction and operation of built assets (WGBC, 2019), 70% comes from the use phase. In addition, a 1% increase in productivity driven by improved working environments can equate to a £20 bn increase in national output (BIFM, 2016). So, not only does it make sense to look at the whole life value of a proposed development or refurbishment, but it is also important to look at how existing assets can be operated more efficiently and provide a productive and supportive environment for their occupants. The carefully considered and planned use of technology and automation can support all these endeavours.

The smart building market alone is expected to be worth in excess of US$100 bn by 2025 (Arcadis, 2022), but this chapter is not just about smart buildings. A smart building is one which has been built or retrofitted to include technology which is capable of gathering and monitoring operational data and is capable of acting on that data to deliver better outcomes. While smart buildings are undoubtedly a worthwhile goal, many of the technologies can be beneficial on their own and are equally applicable to built assets other than buildings. There are also technologies that can help improve the operation and management of multiple assets together, and which can be applied to assets of all scales and sectors. On a domestic scale, for example, video doorbells and security systems have revolutionised home security and made the technology more accessible, while smart thermostats can learn and operate preferred heating patterns. At the other end of the scale, drones and machine learning can be used for preventative maintenance of large infrastructure assets without the need for regular, expensive and potentially dangerous inspections by teams of humans.

The move towards considering the whole life value of an intervention from the very start can only be beneficial, and provides the right foundations for the application of technologies to optimise built assets. While retrofits of technologies can be hugely beneficial, it is undoubtedly better to be able to plan for the use of technologies from the outset. Part of this planning involves thoroughly understanding what the desired outcomes are and how occupants will interface with the technologies, as well as ensuring that information management during the delivery phase supports future operations. It is also important to consider the expected life cycles of technologies relative to the structures they support. Smart technologies continue to develop at significant pace, but the last thing that anyone wants is for a smart asset to become dumb within a couple of years of operation without the ability to make reasonable upgrades, simply because the need for regular improvement was not considered from the outset.

Digital representations of assets form an important part in the application of automation and smart technologies. When dealing with new build assets of a reasonable size and scale, it is common for BIM models to be in place and, in many cases, as-built models will be provided to the asset owner at handover, ready for operation. However, there are clearly many, many instances where this is not the case; when dealing with existing assets, one option to digitise the physical attributes of an asset is to have it scanned.

20.2. Scanning the built environment

Before the development of technology capable of providing accurate, 3D scans of assets, it was the norm to manually measure assets using tape measures and theodolites, and for these to then be drawn up by hand or using 2D CAD. This was usually a very long, laborious process prone to errors, often requiring several trips back to site to check measurements. Still, the end results were often difficult to share between parties and did not necessarily provide accurate representations of the assets they presented, potentially relying on a degree of interpretation. With the development of laser measures, the accuracy of on-site measurements has improved, but still significantly better results can be achieved using laser scans or photogrammetry. These scans can provide the basis for accurate digital representations or digital twins of existing assets which can be used for a number of purposes.

With any form of measured survey, it is important to first identify the level of accuracy that is required – for example, is the purpose of the survey simply to work out floor areas for rental agreements or to measure for installation of precision manufactured equipment? RICS (2014) provide useful guidance on accuracy bands that can be used to identify the level of accuracy required, and this will clearly have an impact on the technology or tools used to carry out a survey.

As described in Chapter 19, 3D scans can be created using cameras attached to helmets, or even with lidar-enabled mobile phones. Such technologies use photogrammetry to combine photographs and create 3D models but, as they are taken with a camera in motion or in many different locations, they are not the most accurate of scans. They are, however, still suitable for many purposes, including progress monitoring or identifying whether objects are present or not. The most accurate forms of scans are laser scans, which are typically taken from equipment fixed to a tripod. These can take scans that are accurate to within a centimetre, and can take full scans of enclosed spaces in less than 30 seconds. Clearly, they can only scan what is visible from the scanner, so it is common to need scanners to be relocated to get multiple angles, and then scans can be combined to create one accurate model. While the scanning is a quick process, equipment tends to be very expensive, and it is always best to use experienced individuals to carry out and process scans to give the best results. Processing is likely to be required not only to combine a number of scans but also to interpret what has been captured and to remove items of furniture, for example, which may be present in a scan but not relevant for the purposes of the survey.

For large outdoor areas in particular, or for surveys of tall assets such as building façades and roofs, drones are now commonplace for producing accurate, quick and safe surveys that can readily be used in 3D modelling. Commercial grade drone scanners can produce point clouds in the same way as tripod-mounted laser scanners but, for safety reasons as well as accuracy, surveys should always be carried out by experienced and qualified pilots.

As technology continues to develop (including mobile phone cameras), it is likely that the use of 3D scans will become more accessible to those working on smaller, domestic developments. However, the scans still need to be processed and useable. There are smartphone apps that can be used to view and even measure from point cloud surveys, but without processing they can be unwieldy and, while they may be useful to enable tradespeople to view assets without having to return to site, it is still likely that the preference is to take measurements manually on site rather than rely on technology.

Scans are more likely to be used by those that are remote from the site in question, or potentially to prove that work has been carried out and for evidence to be uploaded to platforms for others to see.

In addition to forming the basis for digital representations of assets and measured surveys, scans can also be invaluable when it comes to monitoring the condition and performance of existing assets.

20.3. Monitoring the built environment

There are two main purposes for monitoring assets in operation. The first is to monitor the condition of the asset and whether everything is where it should be or to predict future maintenance requirements. The second is to monitor how that asset performs, whether structurally, environmentally or how it is used and whether it remains optimised for the purpose for which it is intended.

20.3.1 Condition monitoring

Condition monitoring of some of the world's largest bridges and infrastructure assets can be very expensive, time consuming and dangerous. However, using Internet of Things (IoT) sensors, drones and AI can have a significant impact on delivering the desired outcomes. Sund & Baelt use drone scans and IBM Maximo Civil Infrastructure to replace the need for teams of mountaineers to inspect bridges (IBM, 2022). Instead, images are captured and the AI (using computer vision and machine learning) rapidly and accurately scans the images, identifying any cracks or damage, which are then assessed more closely. This monitoring process can take a day, whereas the same activities previously took a month or more. Not only that, but structures can be assessed much more regularly. This ability to either monitor in real time or at more regular intervals, and to use AI to analyse results, enables a shift to more preventative than reactive maintenance.

A common approach to managing assets is to use a computer-aided facilities management (CAFM) solution, which can manage maintenance and scheduling, space management and asset tracking and predictive maintenance costs. Common examples include IBM Maximo and QFD by Service Works Global, both of which have solutions for buildings and infrastructure assets.

There are a number of technologies available to assess the structural integrity of concrete and steel products, as well as scanning concrete to locate reinforcement. Screening Eagle (described in Box 20.1) provide a number of solutions, along with an inspection and monitoring platform to speed up the inspection process and sharing of results. Such platforms not only enable data to be produced in a common format but also make it more accessible to those that need it, whether remotely or on site.

Box 20.1 Screening Eagle

Screening Eagle provide a number of technologies for inspecting, managing and maintaining concrete structures and mapping of underground services. Their Inspect software supports visual inspections by locating inspection points, photographs and overlaying sensor data, as well as supporting 3D scanning and providing inspection checklists, while Defect uses a neural network engine to identify and segment cracks from images, and other technologies enable nondestructive concrete assessments, such as corrosion, compressive strength and uniformity and thickness.

Sensors such as Optical Displacement Sensors (ODS) by Senceive can also be used to accurately monitor potential movement in structures, such as vertical movements due to settlement, lateral movements such as rail track slew and earthworks or embankment movements.

Condition monitoring is not restricted to only structural aspects. For example, United Utilities have used a range of technologies to automate the detection of leaks across their network. They have used FIDO AI (described in Box 20.2 and illustrated in Figure 20.1) to identify, locate and even size leaks in underground pipes within their network. To identify leaks over larger areas, they have also trialled drone scans and algorithms from Sensat to remotely map thermal data and photogrammetry, which are then analysed to identify potential leaks.

Box 20.2 FIDO AI by FIDO Tech

Figure 20.1 FIDO AI uses its AI solution to identify and analyse the location and size of leaks (Source: FIDO AI)

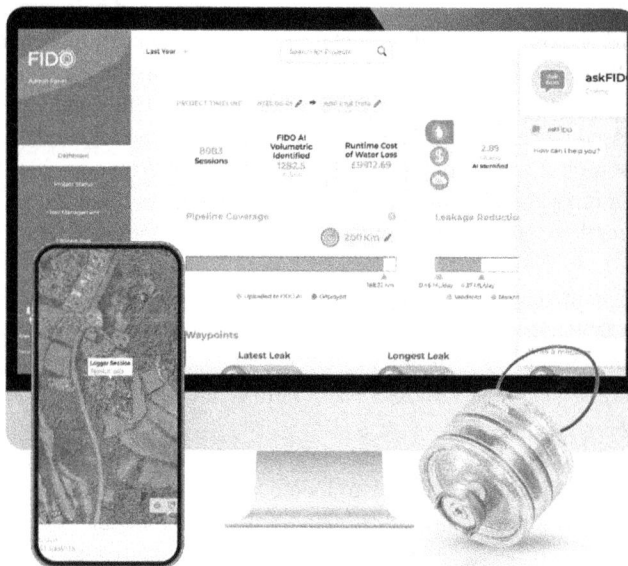

FIDO AI identifies leaks in water pipes by picking up sounds, vibrations and kinetic data. It then uses its AI solution to analyse the location and size of leaks so that they can be prioritised. It can then validate volumetric savings and the effectiveness of repairs. It has delivered over 37% leakage reductions in projects to date.

A simple application of technology that can be useful in the maintenance of built assets is in the form of QR codes printed on products or, where they exist, links to product data from as-built 3D models or digital twins. Such techniques make it significantly easier to access relevant product and warranty data than having to read through hard copy operation and maintenance manuals.

Another potential future use of condition and performance monitoring is to assess how construction products and systems perform in use. Most construction products undergo significant testing to assess and evidence how they perform under defined conditions, often under laboratory conditions – for example, testing the thermal resistance of insulated panels or the structural capacity of precast concrete components to assess compliance with relevant standards. However, those products will often perform very differently in real-world situations. For example, thermal properties can be impacted by neighbouring materials, connections, penetrations and surrounding climatic conditions. While exact performance cannot always be accurately predicted, with the use of sensor technologies and artificial intelligence, confidence in predicted actual performance can be significantly increased. The industry is a long way from certifying products based on their performance in use, but evidence on how they have performed in similar situations will help make more informed decisions when developing solutions. As will be described in Chapter 21, performance monitoring of products and data on the conditions in which they have been used can also prove useful when considering their suitability for reuse.

20.3.2 Performance and use monitoring

The use of movement sensors both within and outside of buildings is nothing new to control lighting, for example, but now that data and much more can be recorded and monitored remotely. Sensors are described in more detail in the next section, but what they provide is potentially real-time monitoring on how assets are performing and how they are being used. Armed with this knowledge, action can then be taken either directly through automation or indirectly with human consideration and input. Envelo (described in Box 20.3) provides a monitoring and analytics solution that focuses on the health of a building's spaces, identifying actions that can be taken to make improvements.

Box 20.3 Envelo

Envelo is a technology company that focuses on the balance between health and wellbeing and energy outcomes. Using IoT sensors, it monitors real-time conditions and occupancies, and uses predictive learning to identify actions that can be taken to reduce risks and improve the health of occupants indoors, such as adjusting heating or air conditioning. In some instances Envelo has reduced winter energy consumption by over 45% and delivered return on investment of over 150% in the first year.

Performance monitoring is not only restricted to energy usage and environmental performance, however. For example, sensors can be used to measure the flow rate and volume of liquid in pipe networks, which can then be used to control or divert flows to other routes to maximise the capacity of a network. Sensors can also be used to measure waste production and collection within buildings, using individual bin sensors to aggregated waste collection points.

Ultimately, the data obtained from monitoring can be used to measure, monitor and either inform or actively change an asset's operations to optimise its suitability and performance. Technologies to automate active control are described in Section 20.5.

20.4. Sensors and other smart measurement technologies

The world is becoming more connected day by day, with the number of IoT connections predicted to increase from 1.7 bn globally in 2020 to 5.9 bn in 2025 (Ericsson, 2022). A significant proportion of these connections will involve the built environment, and its monitoring and operation.

There are many different types of sensors that are relevant to the monitoring and operation of the built environment, and below is a short overview of some of the different types and their uses

- *temperature sensors* – these measure the amount of heat generated from an area or object and can be used to identify the need for heating or cooling
- *proximity sensors* – these detect the presence or absence of people or objects without physical contact, and are commonly used to control lighting
- *pressure sensors* – these detect changes in pressure of a gas or liquid and can be used to identify leaks
- *water quality sensors* – these monitor different aspects of water quality including acidity
- *chemical and gas sensors* – these monitor the presence of toxic or hazardous gas
- *smoke sensors* – used to monitor the presence of smoke
- *level sensors* – commonly used to measure the level of materials or liquids, such as fuels or waste
- *humidity sensors* – used to measure the amount of moisture in the air and are commonly used in heating, ventilation and air conditioning systems (HVAC).

In addition to sensors being capable of measuring performance, it is also important to be able to clearly identify where sensors are located. While this can be done by manually recording a location, and some sensors can also identify their own relative location, there will be times when accurate location systems such as Quantum RTLS (described in Box 19.4) will be valuable, especially when consideration of location in three dimensions is relevant. Sensors measuring performance or structural integrity will typically be static, but location systems may be important to identify the movement of machinery or even users relative to an asset for health and safety purposes.

20.5. Active control and management

Technologies now exist that can take the data from sensors or digital twins and actively control an asset's systems to optimise either use conditions or energy use. On a domestic scale, for example, smart thermostats such as Google's Nest can automatically turn heating on and off, and to different temperatures, based on analysis of occupancy patterns and the location of occupants. Lighting systems can be controlled remotely through a smart phone, security cameras can be viewed and doorbells answered using the same device. On a larger scale whole building or zonal heating, lighting and air conditioning systems can be automated based on performance criteria identified by asset managers and managed through a single interface.

To manage the heating, power, ventilation and air conditioning systems in nondomestic properties, a building management system (BMS) or building automation system (BAS) is often used. In simple terms, both BMS and BAS are computer-based control systems to remotely manage services. An intelligent BMS can bring together control systems and data from a number of different sources and in complex assets and manage them in one interface, whereas a BAS is typically much more basic. Optimise AI (described in Box 20.4) takes the management of systems a step further, by linking a BMS (where one exists) with a digital twin, sensors and artificial intelligence to optimise the energy performance of a building. However, the technology does not rely on having these technologies in the first place, and can start with a simple meter reading.

Using Kiona Edge, an intelligent management system incorporating sensors, artificial intelligence and a BMS, an example property owner saved over 10% of energy costs annually through energy management and optimisation alone (Ericsson, 2022); these savings are not wildly out of character for the level of savings that can be made through the use of such active systems.

> **Box 20.4 Optimise AI**
>
> Optimise AI have developed technologies that combine real-time data from digital twins with artificial intelligence to analyse past performance and predict future performance of nondomestic buildings. Optimise by Optimise AI shows how a building is performing in a live 3D model, and provides recommended actions to achieve savings as well as quantifying what those savings are likely to be. Actuate by Optimise AI goes further and enables artificial intelligence to optimise both energy producing and consuming devices, based on historical performance and user data and real-time climate data, by connecting the building management system (BMS) to smart meters, sensors, actuators and the digital twin.

As described earlier, it is important to first define desired outcomes, and then to find the technology to help deliver them.

20.6.   Technology relevance summary

Using the modelling framework described in Chapter 6, Figure 20.2 summarises the potential relevance of different automation technologies to the use phase of an asset.

Figure 20.2 Technologies relevant to the use phase of assets (Author's own)

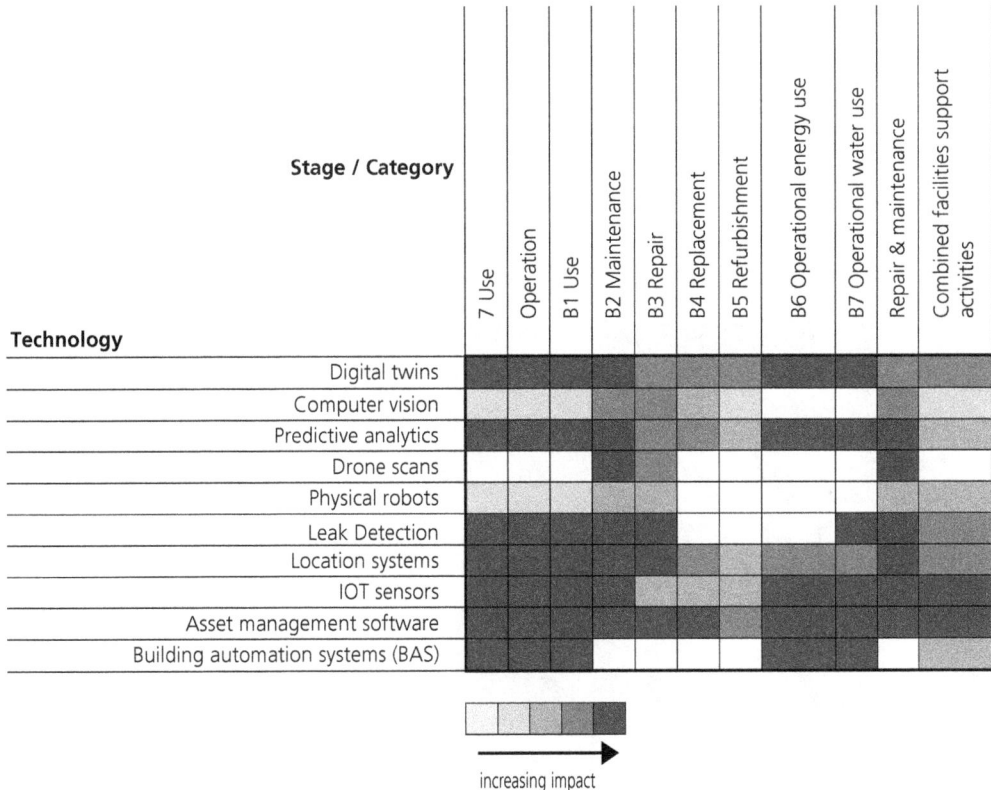

increasing impact

REFERENCES

Arcadis (2022) *The Business Case for Intelligent Buildings.* Arcadis, Amsterdam, The Netherlands.

BIFM (British Institute of Facilities Management) (2016) *The Workplace Advantage.* Raconteur, London, UK.

Ericsson (2022) *Connected Buildings Energy Management*, Ericsson, Stockholm, Sweden.

IBM (2022) Building bridges to better insight. https://www.ibm.com/case-studies/sund-and-baelt (accessed 19/03/2024).

RICS (Royal Institution of Chartered Surveyors) (2014) *Measured Surveys of Land, Buildings and Utilities*, 3rd edn. RICS, London, UK.

WGBC (World Green Building Council) (2019) *Bringing Embodied Carbon Upfront.* WGBC, London, UK.

Section 6

Optimise

emerald PUBLISHING   ice

Steve Thompson
ISBN 978-1-83608-599-7
https://doi.org/10.1108/978-1-83608-598-020241024

Chapter 21
Circularity enablement

21.1. Introduction

Earlier chapters have highlighted the need for different models to deliver, maintain and upgrade the built environment, and a key reason for this is the need to deliver and manage more assets with less resource. The UK's target to achieve 78% lower carbon emissions by 2035 means that, while embodied carbon in 2021 made up 20% of built environment emissions, operational carbon will reduce to the stage where embodied carbon makes up more than half of the total by 2035 (UKGBC, 2021). One clear way of reducing embodied carbon is to optimise the life cycle of assets and their constituent parts, which typically (but not always) means using assets for longer and reusing their products and systems, thus reducing the need for new materials and products. An economy where resources are reused without the need for new materials to be mined and developed is known as a circular economy, and this chapter describes how both manufacturing-led construction and automation can support a move to a circular built environment.

In the UK in 2020, over 92% of all construction waste was recovered and reused or recycled in some form (DEFRA, 2023), which on the face of it sounds impressive. However, the majority of that material was down-cycled to a lower grade use such as backfill, meaning that new energy-intensive products still needed to be produced. Globally, only 7% of products were reused in 2023 across all industries, down from 9% in 2016 (Circular Economy Foundation, 2024). In the Dutch construction market the figures are similar, with 88% of materials being reclaimed but only 8% of products being reused (Circular Economy Foundation, 2022). So, there is a need to both reduce the amount of waste and increase the percentage of any waste that can be reused and not just down-cycled. The Circular Economy Foundation (2024) describe the need to develop strategies that

- narrow the material footprint by *using less* resources
- slow material flows by extending or optimising the life cycle of assets and *using them for longer*
- cycle materials by *reusing* products, systems and materials.

Figure 21.1 illustrates the stages of an intervention, and the potential flow of products and materials from assets back into the supply chain through reconfiguring existing assets or reusing, redistributing, remanufacturing or recycling products.

To optimise the use of products, materials and assets over their life cycle, it is first important to move from perceiving a building or structure as a finished, static product from the point it is handed over, to seeing it as an evolving asset which is capable of accommodating changing user and community needs. This can be a big shift in mindset, but one that needs to be made in order to reduce the resources used through an asset's life cycle. Approximately 80% of the building stock that will exist in the UK in 2050 already exists (UKGBC, 2021) and so a significant number of assets will need to be either replaced, upgraded or adapted to deliver the UK's carbon reduction targets and

Figure 21.1 Flow of products and materials into and out of assets and back into the supply chain (Author's own)

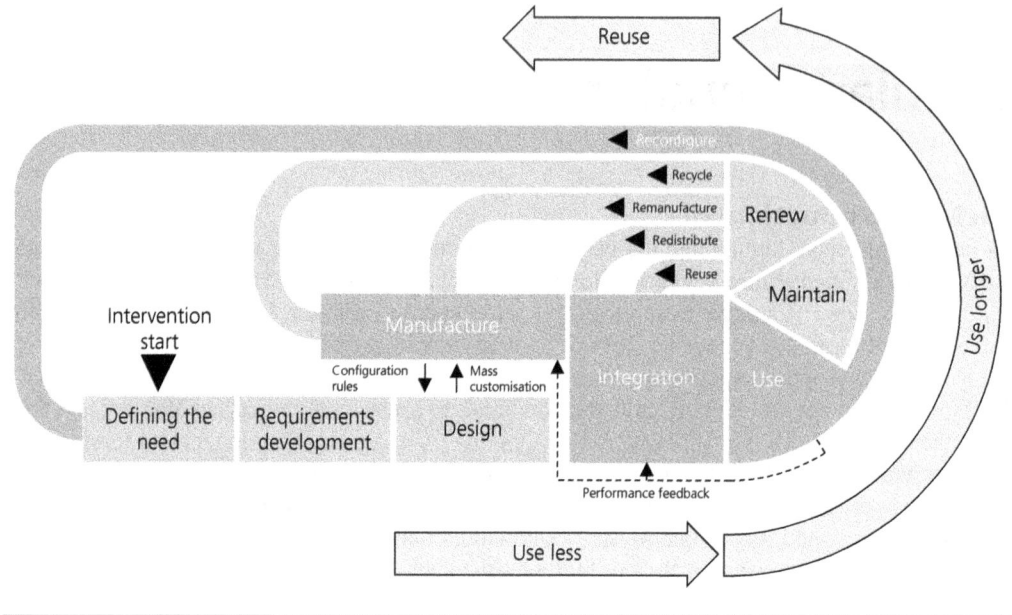

ensure that they remain suitable for changing requirements. That means that a significant amount of work needs to be done that involves existing products and systems, often replacing them with new virgin products. It is important to consider how those products being replaced can best be utilised, but also to ensure that new assets built in the future are designed to consider future reuse of materials, products, systems and the assets as a whole. Design for manufacture and assembly (DFMA) alone is not enough; delivering new assets with systems that cannot easily be reused or reconfigured only further exacerbates the problem. There are significant opportunities going forward to use a manufacturing-led mindset to deliver and maintain built assets to achieve more sustainable solutions but, to be successful, the whole life cycle of assets, products and systems need to be considered together.

The shift from static to evolving assets begins with designing for new assets (or modifications to existing assets) to be delivered and maintained with fewer resources.

21.2. Use less

Designing static assets with the aim of minimising the amount of material used can mean designing less or smaller spaces that are optimised to meet a set of well-defined requirements. However, this is rarely advisable as it is likely to be incredibly restrictive and mean future adaptability is virtually impossible. There is a phrase sometimes used in reversible building design (buildings that can be transformed or dismantled without damage), which is that any waste is a mistake. That can be true, but only when the full life cycle and required flexibility and adaptability of an asset is considered. Designing in future flexibility is not a waste, or a mistake, but designing with no consideration for material optimisation or adaptability can be inadvisable and leads to unnecessary resource use. Structures, for example, can be optimised to use the least amount of material to deliver solutions that can be reconfigured or disassembled and reused. Designing in the ability for products and systems to be dismantled and reused, and for spaces to be reconfigurable, is likely to save material use in the long run, as materials do not need to be wasted, down-cycled or replaced with new material. Designing with such capability to reconfigure or upgrade when using manufacturing-led solutions can lead to industrialised built environments as described in Chapters 4 and 14, and the precision that such approaches bring can significantly minimise waste in the process.

21.3. Use for longer

There are two main reasons for remodelling or replacing an asset. The first is to provide a different spatial configuration (which can be for the same primary function, such as cellular to open plan offices, or changing from one function to another, such as commercial to residential, or to expand or contract an asset to provide more or less space). The second reason is to improve the performance of an asset, whether to reduce energy consumption or provide a different internal environment. Where an asset already exists, such modifications should aim to minimise damage to existing products and systems that can potentially be reused, and avoid abortive work. The need for further flexibility in the future also still needs to be considered, such as performance updates described in Section 21.3.2.

Where a new asset is proposed, again future flexibility in terms of function, configuration and performance will ideally be considered from the outset. Such an approach will not always cope with every eventuality but, in developing user requirements, consideration of potential future likely scenarios can minimise the risk of unnecessary wastage down the line. When taking a longer-term perspective than is traditionally the case, manufacturing-led solutions are likely to become even more appropriate than just the effective delivery of static assets due to their precision, performance certainty and suitability for reconfiguration and reuse.

21.3.1 Reconfiguration

There are times where it makes more sense to design an asset for a short life cycle instead of designing in unreasonable amounts of flexibility that makes the solution inefficient. In such scenarios, the asset will ideally be designed so that its constituent parts can be readily dismantled and reused, with the aim of replacing it with an asset developed with a similar philosophy. An example of where an asset can benefit from being designed for a short lifespan and reuse is event structures such as temporary stadia, and an example of assets where a short initial solution needs to be adaptable to change to a longer-term solution is athletes' villages for sporting events such as the Olympics. In such scenarios, the initial use as single living accommodation is likely to make way to either guest or family accommodation for the longer term, and these facilities should be designed to minimise wastage in the transition between different use scenarios.

While short life cycles are occasionally preferable or required, generally it is better to design assets to have as long a life cycle use as possible, which means that user requirements are likely to change and spaces need to adapt over that period. In preparing for future reconfigurations, it is important to understand common space types and characteristics, but also building service requirements and adaptability processes. For such scenarios, it is worth spending time clearly defining needs of different scenarios as described in Chapter 11, but also considering product platforms for solutions that fit multiple use cases. Solutions such as the ADPT circular building system by Futur2K and Arup (Arup, 2024) are designed to not only be reversible and scalable to suit different life cycle scenarios, but are also designed to be optimised for different component life cycles.

While it is always preferable to consider future asset configurations and requirements at their initial development stages, when dealing with existing assets, clearly that is not possible. That does not mean that reconfigurations are not possible, it just provides a different starting point. It may be that internal alterations can be applied in such a way to enable further future reconfigurations (for example, using demountable and relocatable partitions and flexible service strategies), or potentially extensions are designed to offer future adaptability, breathing new life into existing assets. To be successful, such solutions need to go way beyond just the addition of extra space and encompass service strategies and potentially demountability of components without waste or damage, and manufacturing-led solutions are ideal to provide this flexibility.

Reconfiguration is less likely to be required or possible with linear assets or structures such as bridges or tunnels, as these are typically designed for one specific purpose that is unlikely to change. However, there are examples, including the Castlefield Viaduct in Manchester, which has been converted into a park.

21.3.2 Performance updates

Performance upgrades of our built assets are essential if the UK is to meet its zero carbon targets in the next two decades. Heat pumps are expected to replace gas boilers in over 600 000 homes per year by 2028 (BEIS, 2023), but for them to operate efficiently many of those homes will first need to be better insulated. Solutions such as those delivered by Energiesprong and described in Box 14.3 in Chapter 14 are already using manufacturing-led solutions to deliver zero carbon homes. Solutions such as Ventive Home provide performance upgrades to heating, ventilation and hot water systems while minimising disruption to existing facilities. A new 'box' can sit inside or outside of a home and connects to existing systems, so can easily be upgraded in the future without further disruption to the existing building. Ventive also provide solutions for nondomestic assets, where passive ventilation with heat recovery (PVHR™) reduces heat loss by up to 72%.

To improve the insulation of existing assets, prefabricated panels can provide quick, accurate and high-performance solutions that can be used on assets that are otherwise difficult to treat. KMT Prefab, for example, start by carrying out laser scans of the existing asset and then deliver prefabricated overcladding solutions for the upgrades.

The move towards consideration of the life cycle of assets and the ability to track live performance can provide opportunities for product and system manufacturers to develop new business models including a performance as a service model. Such a model can be used to guarantee performance (for example, guaranteed internal temperatures or ventilation) and mean that the supplier can perform upgrades and reuse components while providing the agreed performance to users. An example of a performance as a service approach is Octopus Energy's Octopus Zero Bills, where home users are guaranteed to have zero energy bills for five years as a result of having a home with certain components installed, including solar panels, batteries and heat pumps. While this is currently focused on new-build homes, the initiative plans to move towards retrofits that meet the specifications in the future.

Unlike buildings, structures such as bridges, tunnels or horizontal assets are less likely to be retrofitted for performance upgrades as their performance requirements are more likely to be purely structural than environmental, and it is more likely that components are customised for a specific application. Bridges can potentially be repurposed, for example, from a road bridge to a pedestrian bridge carrying less load, but that does not reduce the need for a replacement bridge to be constructed. However, it is more likely in the future that, with the combination of an increased focus on the circular economy, availability of monitoring technologies and availability of data on performance in use, structural assets will be designed to be dismantled and reused either as a whole or as individual components.

When planning for future maintenance or upgrades to assets, it is important to consider the service life of what is being constructed, and the relationships between elements with different service lives. For example, if services are placed behind wall structures with a considerably longer service life, the wall structure will need to be removed or damaged in order to maintain or replace the services, leading to unnecessary work unless they are designed in such a way as to be removable and reused for such purposes.

21.3.3 Service life planning

The service lives of assets and their constituent parts are affected by a number of different factors, including connections between products and materials, the quality of materials used, the environment within which products operate and the level of care and skill in installing products. An understanding of the service lives of products is relevant to understanding future maintenance (and replacement) works, to assess the future value of products and systems, and the potential and availability for reuse, as well as to ensuring that abortive work is designed out by configuring components to minimise unnecessary penetrations. A useful reference on the service life of components is the BS ISO 15686 (BSI, 2017) series of standards. This includes the factor method, which is a method to estimate the potential service life of a product by considering factors that will impact its life when compared to conditions under which a reference service life (RSL) has been calculated. The categories of factors include the inherent performance level of the product, the level of protection it has, the skill and care taken in installation, likely wear and tear in use, the severity of both internal and external environments, and the level of maintenance that is proposed.

Such an approach relies on having a clearly defined RSL, which will typically be assessed under factory conditions and is not always available. However, the increasing availability and use of sensors and artificial intelligence provide new opportunities to estimate service lives based on performance in use under similar conditions.

When developing manufacturing-led solutions, there is a real opportunity to optimise service lives of products and systems, but to enable this to happen the issue needs to be considered from the outset. This will have an impact on how products are connected, and potentially their position within a layered solution. Using manufacturing-led solutions also improves the potential for reuse of components and systems due to their precision and potential for alternative connection systems that potentially suit disassembly without abortive work.

21.4. Reuse

Once the life cycle of an asset has been optimised, it is important to consider how components and systems can be reused with the minimal amount of effort and waste. There are two main aspects to consider, which are designing to enable disassembly with minimum disruption and waste, and the effort that is required to enable the product to re-enter the construction supply chain and be reused.

21.4.1   Design for disassembly

Disassembly is the nondestructive taking apart of assembled products or systems into their constituent parts. It is different from demolition and component recovery, as that is likely to involve destruction of some elements in order to recover others, and is likely to be carried out as an afterthought, not part of a planned process for extracting and reusing products.

If design for disassembly is considered from the outset, it can enable alternative business models to the traditional sale model, such as renting of building systems or manufacturer's take-back where suppliers can either take back for free, or buy back products at the end of their use cycle. When it comes to disassembly, it is also worth considering the degree to which elements are best disassembled. For example, many construction components use a combination of materials (such as steel and insulation for cladding panels), and it may be more advantageous to retain the composition as panels instead of splitting down to the material level, as it retains the product's use at a higher value. The optimum level will depend on issues such as connectivity, tolerances, product edges and of course the relative life cycle of components making up a product. Where possible, however, it is best to support more than one reuse option to increase the likelihood of it occurring.

Industrialised construction can offer a significant advantage in promoting disassembly as it typically uses standardised components or interfaces, and uses more bolted rather than chemically bonded connections, with long-term partnerships and coordinated supply chains that are more likely to benefit from and be capable of delivering product reuse with real knowledge of what the product is and its capabilities. Box 21.1 describes a system by Natural Building Systems, which is not only designed to be flexible and to capture carbon, but also to be dismantled and its components reused. Another common application of disassembly is in the rental of temporary buildings, where structural frames or even entire buildings can be rented out and fitted out for different purposes, and the frames can then be reused in future applications.

When considering the potential future dismantling of an asset, it is important to consider what information needs to be retained to make the effective reuse of materials as easy as possible. For these purposes, materials passports and other information management tools are invaluable, and are described in Section 21.5.

21.4.2   Re-entering the supply chain

Whether an asset has been designed with reuse of products and systems in mind or not, there comes a stage when products need to be removed in some form and go through a series of processes to re-enter the supply chain. A simple way to consider the steps required and suitability for products to be reused is to use the Five Ps model described in Chapter 3. It works the same way as for virgin products entering the market, but in this case the starting product is one that has already been used and may need an element of reworking (for example, bricks will need to be dismantled, have any mortar removed and cleaned before they can be reused). This approach will help understand the potential work and cost involved, as well as the most appropriate part of the supply chain for a product to re-enter.

A potentially significant area that needs to be considered when reusing products is testing and certification, which should not be underestimated. Especially when it comes to the reuse of structural components, there needs to be confidence that the product being reused has the required properties for it to be suitable. With the latest iteration of the Construction Product Regulations (CPR)

Box 21.1 Natural Building Systems

Natural Building Systems have developed a fully integrated, prefabricated kit of parts system including high-performance breathable walls, roof and floor components which are designed to be demountable (Figure 21.2). The panels are connected using patented removable timber pegs with tongue and groove cones, which can enable rapid assembly and disassembly and make it easy to completely reuse elements or make nondestructive alterations.

The system can be delivered as panels or volumetric units, and individual panels can be removed to allow alterations and access for upgrading services, such as plumbing and wiring. The system is a combination of engineered timber and hempsil, a bio-composite, and is net carbon negative, meaning the system removes more carbon from the atmosphere than it takes to make it. Digital tools are used to design and configure the kit of parts for individual projects, and each component includes a digital product passport, sharing important data which supports future maintenance, dismantling and reuse.

Figure 21.2 Natural Building Systems have developed a fully integrated, prefabricated kit of parts system which is designed to be demountable and its components reused (Source: Natural Building Systems)

(HMG, 2013), products re-entering the market in Europe will be treated the same as new products, in that they can be CE marked when tested.

For the reuse of structural steel, the Steel Construction Institute (SCI) provide guidance on the requirements for testing and designing with reclaimed sections (SCI, 2019), which requires a factor to be applied to the design of reclaimed steel to provide comfort that the sections will be suitable for their new purpose. FerrousWheel by Symmetrys (Symmetrys, 2024) is a tool that automates the process of matching designed steel solutions in Autodesk Revit with reclaimed sections.

Reinforced concrete is another widely used structural product in the construction industry, but traditionally is demolished at the end of life. However, it can be reused; Küpfer *et al.* (2024) have developed solutions to enable reinforced slabs that were cast in place to be cut from donor buildings, and reused as short span slabs in new buildings. Such an approach can significantly reduce waste and embodied carbon and relies on tools and plant that are regularly used within the construction sector.

The potential for reuse varies significantly depending on the product or system in question, but also on awareness of what products are available, when and where. Where a product or system manufacturer is involved in the reuse in the form of take-back or reuse, then the process should be relatively straightforward in that the supply chain is already in place. The manufacturer and original supply chain will not always be involved, however, and under those circumstances platforms such as Enviromate and Site-Spares offer marketplaces for the sale of spare products, while Salvo focuses on the sale of reclaimed products. However, each of these solutions relies on the user first deciding to use reclaimed products instead of new. To significantly increase the use of reclaimed products, it is likely that system providers (for example, building system or industrialised construction providers) will need to become more involved, which is more likely to become a reality with the combined use of manufacturing-led construction and digitalisation where reclaimed products are available for consideration alongside new.

21.5. Information management

There is no question that the provision of detailed and reliable digital information on a product can be invaluable in understanding what the product consists of, what previous applications have been and potentially what conditions the product has been exposed to. This information is crucial to move towards a circular construction economy and to enable suitable reclaimed products to become visible to the market. It does not mean a central database of all products and systems, rather that the owners of buildings will have a comprehensive record of what their assets contain, and what can be reclaimed and how. Material passports provide very useful data for this purpose, and templates for these are freely available (Material Passport, 2024); digital product passports are a requirement of the new Construction Product Regulations from 2024. Material passports and digital product passports are both discussed in Chapter 10.

The Reconmatic (Reconmatic, 2024) initiative was launched in 2023, and aims to use BIM and data to improve the construction industry's waste management. As a part of this project, a new data set to share construction waste data is being developed, known as WASTEie, which is based on the IFC schema. The purpose is to make the sharing of data on waste more consistent and easier to exchange between parties, which will make waste management and reclamation easier in the future. This is particularly useful where products are not reclaimed and reused. Finally, the bSDD highlighted in Chapter 10 includes a decommissioning and reuse data dictionary to identify properties to support the reusability of products and materials.

21.6. Conclusion

Especially as the construction industry shifts its focus from the delivery phase to an asset's whole life, there is more reason than ever to look at the whole life cycle of its constituent parts, how their life cycles can be optimised and how assets, spaces, products and materials can be reused to minimise their environmental impact and deliver a sustainable built environment. The move to a more manufacturing-led approach supports this by enabling the dismantling and reuse of systems and components, ready for re-entry into the market. A systems approach makes it easier due to the accuracy of components, standardisation across assets and the fact that supply chains to reuse components are already in place, with a clear understanding of what is required to bring them back into a useable state. Combined with digital tools that can make such solutions visible (including the clear definition of need and similarities and differences between assets) and automate their design application, the chances for products to be reused are considerably improved over traditional construction mechanisms.

There are a number of system providers who have actively considered reuse in the development of their systems, or how their systems enable the reuse or reconfiguration of spaces, some of which are described in this chapter and elsewhere in the book. To make circularity in the built environment a widespread reality, a systems approach is crucial, with the business model for system reuse being just as important as the technical applicability, and there is evidence that such models are beginning to be used in the market. With organisations such as Energiesprong (described in Box 14.3 in Chapter 14), it can be seen how the life cycle of existing assets can also be prolonged using manufacturing-led approaches, so that the physical and digital technologies are available to extend the lives of many more assets, and a circular industrialised construction sector may be achievable in the decades to come. To make it a reality, it is useful to follow the process described in Figure 1.2 in Chapter 1, which is to enable, define, systemise, automate and then optimise. Without the initial steps, it is unlikely that optimisation will be possible.

REFERENCES

Arup (2024) *ADPT circular building system – modular system for circular buildings*, Arup. https://www.arup.com/projects/adpt-circular-building-system (accessed 22/03/2024).

BEIS (Department for Business, Energy and Industrial Strategy) (2023) Energy Security Bill factsheet: Low-Carbon Heat Scheme. https://www.gov.uk/government/publications/energy-security-bill-factsheets/energy-security-bill-factsheet-low-carbon-heat-scheme (accessed 23/03/2024).

BSI (2017) BS ISO 15686: Buildings and constructed assets. Service life planning. BSI, London, UK.

Circular Economy Foundation (2022) *The Circularity Gap Report – Built Environment*. Circular Economy Foundation, Amsterdam, Netherlands.

Circular Economy Foundation (2024) *The Circularity Gap Report 2024*. Circular Economy Foundation, Amsterdam, Netherlands.

DEFRA (Department for Environment Food and Rural Affairs) (2023) Table 5: Recovery Rate from Non-hazardous Construction and Demolition Waste, UK, 2010–2020. https://www.gov.uk/government/statistical-data-sets/env23-uk-waste-data-and-management (accessed 22/03/2024).

HMG (Her Majesty's Government) (2013) The Construction Products Regulations 2013. The Stationery Office, London, UK, Statutory Instrument 2013 No.1387.

Küpfer C, Bertola N and Fivet C (2024) Reuse of cut concrete slabs in new buildings for circular ultra-low carbon floor designs. *Journal of Cleaner Production* **448(2024)**: 141566.

Material Passport (2024) Open Data Format Is the Passport's Digital Language. https://material-pass.org/download/#Download (accessed 15/04/2024).

Reconmatic (2024) https://www.reconmatic.eu (accessed 01/04/2024).

SCI (Steel Construction Institute) (2019) *Protocol for Reusing Structural Steel.* Steel Construction Institute, Ascot, UK.

Symmetrys (2024) FerrousWheel. https://symmetrys.com/climate-response/ferrouswheel/ (accessed 15/04/2024).

UKGBC (UK Green Building Council) (2021) *Net Zero Whole Life Carbon Roadmap, A Pathway to Net Zero for the UK Built Environment.* UKGBC, London, UK.

emerald PUBLISHING   ice Publishing

Steve Thompson
ISBN 978-1-83608-599-7
https://doi.org/10.1108/978-1-83608-598-020241025

Chapter 22
The future of work

22.1. Introduction

This book has covered a range of different technologies and methodologies that can potentially automate and systemise parts of asset delivery and operation. This chapter describes the potential impact of those technologies and approaches on the future of work within the built environment, along with other trends that are likely to have an influence. The basis for the predictions of impacts of automation and systemisation is primary research carried out for this book, the methodology for which is outlined in the Appendix. It uses a bottom-up approach to assess the impact of the 37 technologies described in Chapter 15, along with systemisation as described in Chapter 14. The potential impact of these are assessed against over 1400 tasks identified as being carried out across 66 occupations within the wider built environment, and the results of this work are presented in this chapter and Chapter 23.

It is important to note that automation of existing tasks will have its limits, and the analysis behind this chapter helps determine what those limits are likely to be. Referring back to Figure 1.1 in Chapter 1, there is a difference between doing things right (for example, automating existing tasks) and doing the right things (identifying and carrying out tasks that best take advantage of a combination of automation, manufacturing-led construction and human input).

Automation of tasks is rarely, nor should it be, driven by a desire to replace the jobs of humans. The construction industry in particular is already facing significant labour shortages, exacerbated by the number of experienced workers looking to retire and with the reduction in foreign workers as a result of Brexit. It is likely that, over the next decade, the UK construction industry will lose over 500 000 workers through retirement alone (Make UK, 2023). In a global survey of business leaders across many industries (WEF, 2023), respondents identified key areas of labour shortages, and those most commonly identified were related to construction: brick layers, carpenters, plumbers and concrete workers, followed closely by labourers and civil engineers. At the same time, the industry needs to deliver more than ever before over the next few decades to meet demand and carbon reduction targets. In total it is likely that the UK industry will need to recruit over 950 000 workers by 2030 to meet demand (Make UK, 2023), unless supported by other approaches such as automation and systemisation. They can support by automating suitable tasks and enabling humans to focus on tasks which are either desirable to be, or need to be, carried out by humans. The World Economic Forum (WEF, 2023) suggest that most technologies are likely to increase the number of jobs between now and 2027, not reduce them. The WEF also forecast that in the infrastructure industries (including construction), labour churn (which is the pace of reallocation of jobs) is likely to be 22% by 2027 which, as described in Section 22.2, is in line with the analysis presented here.

The regular WEF report on the future of jobs (WEF, 2023) identifies that of those businesses surveyed, on-the-job training and automation are the most likely strategies to deal with labour shortages, and the three megatrends most likely to have a significant impact on business transformation in construction are the use of new technologies, environmental, social governance (ESG) standards and the increasing breadth of digital access.

The WEF report suggests that 34% of all business-related tasks are already automated. It must be pointed out that this figure is across a number of different industries, many of which include jobs that are potentially easier to automate than in construction (for example, both design and physical jobs in unique site conditions are likely to be harder to automate). Also, those responding to the survey were typically from large organisations, whereas in construction the majority of businesses are very small. Analysis would suggest that the current figure for automated tasks is more likely to be in the region of 10% for construction.

It is important to recognise that automation and systemisation can deliver improvements beyond purely increases in efficiency. For example, location systems do not automate existing tasks directly, but they enable other services to be automated, such as the reporting of progress and location of maintenance to be carried out. Automation can also deliver insights that would not otherwise be available (such as location and size of water leaks or energy performance), improve safety and reduce energy use. However, the focus of this chapter is on the implications for the future of work.

22.2. Future of work assessment

Using the assessment methodology described in the Appendix, the average percentage of time taken to carry out activities related to the design, delivery and maintenance of built assets that can be automated is likely to rise to between 30 and 36%, with between 27 and 32% of tasks being automated, depending on which of the three scenarios described in Chapter 23 are most closely experienced. However, as illustrated in Figure 22.1, the percentage of tasks that can be automated within individual occupations varies significantly up to 93% automation of some occupations' tasks by time.

For this analysis, the percentage adoption of each technology is based on the levels of adoption over time identified in Figure 15.1. The level of adoption of different levels of manufacturing-led construction starts with a total of 11% adoption by value identified by Glenigan (Construction News, 2024) and the relevant proportions of 2D, 3D and subassemblies utilised. The growth in adoption of systemisation then varies by scenario, and is described in Chapter 23.

Figure 22.1 illustrates the potential automation of time by occupation of different skill levels. The black line indicates the average level of automation for tasks, and the bars the total range.

The skills levels are those forming part of the Standard Occupational Classification (ONS, 2010). The skill levels are defined based on the duration of training or work experience required to perform the relevant tasks. They can be summarised as

- Level 1 – requiring a general education, such as labourers, couriers and janitors
- Level 2 – requiring a similar level of education as Level 1, but more work-related training or experience, such as crane operators, drivers or shipping clerks
- Level 3 – requires significant post-compulsory education but not at degree level, such as procurement clerks, carpenters, brick layers or engineering technicians
- Level 4 – typically relates to professional or high-level managerial occupations requiring degrees or equivalent work experience, such as architects, construction managers, civil engineers and cost consultants.

Figure 22.1 Automation of tasks by skill level of occupation (Author's own)

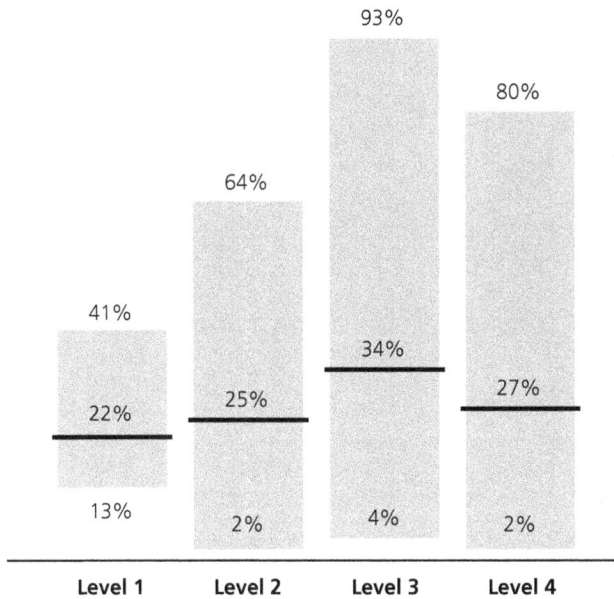

	41%	64%	93%	80%
	22%	25%	34%	27%
	13%	2%	4%	2%
	Level 1	Level 2	Level 3	Level 4

The levels of automation indicated in Figure 22.1 suggest that all levels of roles can benefit from automation, but actually it is likely to be some of the more manual or assistance tasks at level 1 that are harder to automate in construction. With automation and systemisation, it is still unlikely that many tasks will be completely automated – for example, brick laying robots tend to need humans to install brick ties, and design automation needs designers to select and verify results, and so humans will remain involved in the process.

The analysis shows that automation or systemisation is likely to have an impact throughout the life cycle of an asset, and the relative levels of automation can be seen in Figure 22.2. The highest levels of automation understandably relate to manufacturing and integration, as these stages are those most impacted by systemisation as well as automation. Automation alone is highest in the design and procurement areas.

The levels of automation identified in the analysis vary between scenarios, but for the middle scenario, by 2035 it is likely that the level of full-time equivalents (FTEs) released through automation and manufacturing-led construction can be reallocated to grow construction output by in excess of £31 billion. This is another clear incentive to not look at automation as a means of reducing head count, but to focus human resource to work alongside it do deliver more value. These figures are based on the current productivity of construction and related industries as relevant to each occupation assessed, and are considered conservative as they do not assume that they will be reassigned to more productive roles.

The potential for a task to be automated varies considerably depending on the type of task in question. Figure 22.3 illustrates the percentage of tasks of different types with a greater than 50% chance of being automated.

Figure 22.2 Relative automation of tasks by asset life cycle stage (Author's own)

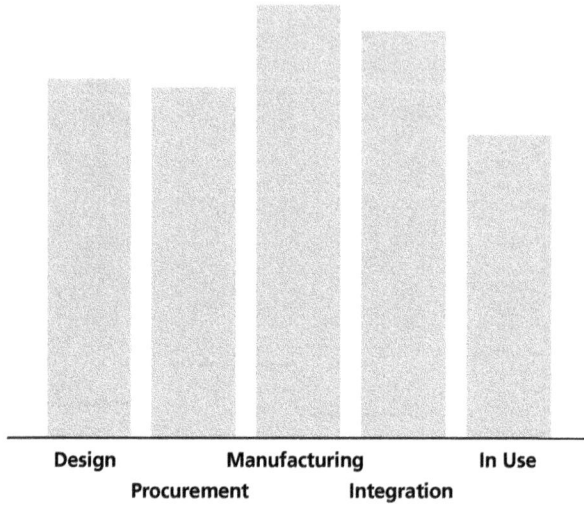

| Design | Manufacturing | In Use |
| Procurement | Integration | |

Figure 22.3 Percentage of tasks with greater than 50% automation by type of task (Author's own)

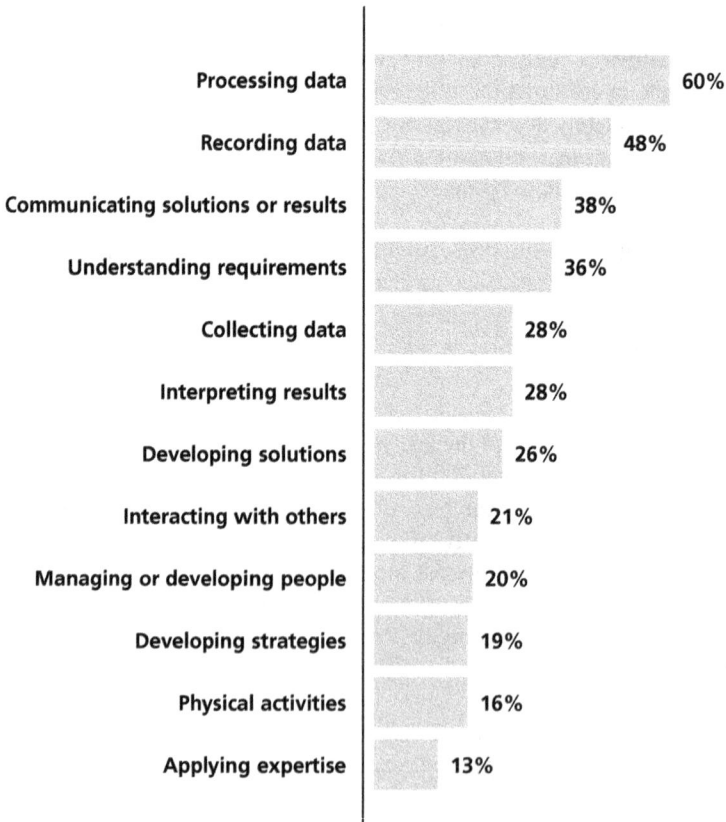

Processing data	60%
Recording data	48%
Communicating solutions or results	38%
Understanding requirements	36%
Collecting data	28%
Interpreting results	28%
Developing solutions	26%
Interacting with others	21%
Managing or developing people	20%
Developing strategies	19%
Physical activities	16%
Applying expertise	13%

This clearly shows that tasks that require the application of knowledge are considerably less likely to be automated than data recording or processing tasks. It is worth noting also that physical tasks are less likely to be automated than nonphysical tasks.

Each of the scenarios described in Chapter 23 are based on applying a combination of automation technologies together. Some technologies do not function without the aid of others (for example, connected autonomous plant without IoT sensors or location systems, and machine learning without a technology to benefit from that learning), and so it can be dangerous to look at the impact of individual technologies in isolation. However, for illustrative purposes of the impact on automation of current tasks, Figure 22.4 presents the relative impact of the technologies identified in Chapter 15.

Figure 22.4 Relative impact technologies are likely to have on the automation of existing tasks (Author's own)

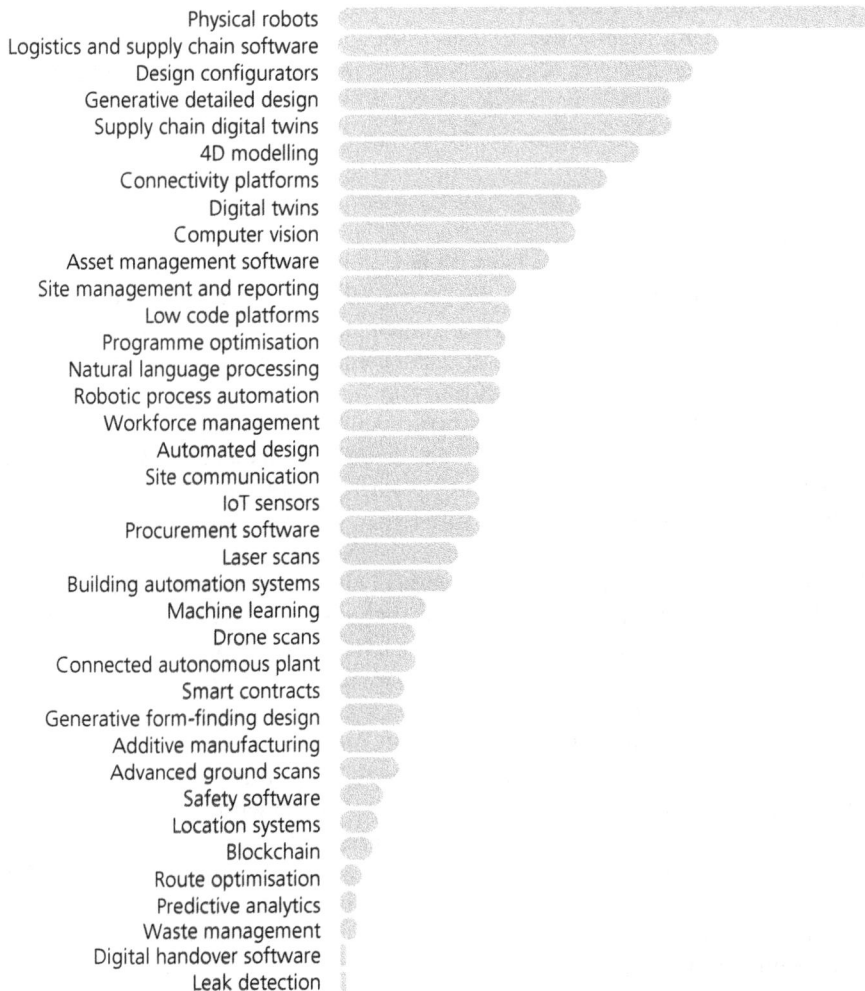

It is important to emphasise again that automation delivers value way beyond the automation of tasks, and so it is important not to underestimate the value that can be delivered by the technologies identified in Figure 22.4. The results presented also relate to the built environment industries as a whole throughout the asset life cycle; technologies such as leak detection clearly represent very significant value in the operation and maintenance of the water sector but not so much in other sectors such as new build residential. Digital handover software too offers significant value, but only relates to a certain stage in the asset life cycle, so when compared to other technologies that can be used throughout the life cycle it appears lower on the list of technologies that can deliver automation.

It is worth noting that in Figure 22.3, physical tasks are among the least likely to be automated, whereas in Figure 22.4 physical robots are shown to have the largest potential impact of any technology. The reason for this is the sheer volume of physical tasks within the construction industry, meaning that with even a relatively small chance of tasks being automated, they can have a significant impact.

22.3. Occupational automation examples

As mentioned earlier in this chapter, overall tasks relating to 66 occupations were analysed, based on occupations which the Office for National Statistics reported were employed in the construction and related industries in 2021 in the UK (ONS, 2023). This section describes three of those occupations in more detail to highlight potential automation of tasks that they are otherwise likely to carry out.

22.3.1 Architects

The detailed analysis carried out suggests that the tasks traditionally associated with architects are among the most likely to be automated by the technologies described in this book over the next decade. To some this may come as a surprise, as architecture is very much a knowledge-based occupation with every project requiring slightly, or sometimes very, different solutions. However, with machine learning, generative design of different forms, smart contracts and detailed laser scans, combined with industrialisation of construction delivery, there are real opportunities to automate tasks. For the purposes of the analysis, it is also important to point out that roles are separated by occupation, so where an architect also carries out other duties such as project administrator and surveyor, the likelihood of automation will be different.

The analysis of current tasks suggests that up to 80% of tasks can be automated by 2035, with the majority of that potentially being possible by 2030. That may seem a very quick shift, but actually many of the traditional tasks carried out by architects have already changed as a result of technology, moving from hand drawings to 2D, then 3D CAD and on to BIM and generative design tools. Clearly, not all architectural projects will be automated to the same degree, partly because of the availability of detailed generative design tools but also down to costs, and the necessary training and confidence to use such tools. However, in some cases, the role of an architect is likely to change more than the 80% mentioned. As has already been described, this shift certainly does not mean that 80% of architects will no longer be required. What it means is that their roles will change dramatically in some cases, enabling them to focus on utilising their core skills. This may mean that on some projects their role changes very little as the technology may not be suited to all projects (or commercially viable). On other projects, however, it may mean that their role is much more about inputting and selecting the right parameters to be entered into generative design tools, and tweaking or selecting the best results from those created. It will certainly still involve the many human aspects of architecture, such as consulting with clients to define or understand requirements and to discuss designs as they develop, but also consultation with other stakeholders throughout the process. The architect may also carry out roles that are not directly related to architecture, such as principal designer, project administrator, project manager or surveyor, in which case the likelihood of automation is likely to change.

The technologies most likely to automate architectural tasks are generative detailed design tools, design configurators, automated design and digital handover software. Where the architect carries out other roles, technologies such as smart contracts and programme optimisation tools are likely to become particularly relevant.

22.3.2 Civil engineers

While civil engineers in some ways hold a similar position to architects in that they can lead the design of assets, the potential for automation of tasks is very different. This is likely due to the complex nature of sites potentially stretching over large distances and the potentially bespoke solutions that are required. The analysis suggests an automation potential of only 29% by 2035 for the tasks identified; however, it must be recognised that some projects will have significantly greater potential than this. An example is described in Chapter 17, where Laing O'Rourke have developed a digital bridge configurator based on a kit of parts approach.

The technologies most likely to automate civil engineering tasks are generative detailed design tools, design configurators, drone scans, computer vision and route optimisation. As with architects, where civil engineers carry out other roles such as project administrator, technologies such as smart contracts and programme optimisation tools are likely to become particularly relevant.

22.3.3 Brick layers

Brick laying is one of the occupations where there is the greatest potential for tasks to be automated, with potentially over 85% of tasks being automated by 2035. This is partly because of the development of brick laying robots (Hadrian X described in Box 19.5 in Chapter 19 is a good solution; however, not all brick laying robots are as refined and can be very limited in their practical use) but also because of the move to alternative methods of construction. Again, the automation of brick laying does not mean that human brick layers are no longer required, far from it. Even the best automated brick laying solutions are not capable of repairing existing assets, accessing internal walls that are under cover or carrying out certain wall types. It is likely that if the analysis is correct, brick layers will be used more on these types of projects that cannot be automated, or projects of a scale where it is not commercially viable to use robots. Another limitation of brick laying robots is that they tend to require other trades around them to be halted during operation, whereas with human brick layers this is usually not necessary, meaning work can continue in parallel. That may not be an issue for large, new-build projects, but for others can be a challenge.

The technologies most likely to automate or replace the need for brick laying tasks are physical robots and additive manufacturing (where the need for bricks is likely to be eliminated), but equally important is the shift towards manufacturing-led solutions, where either 2D or 3D panels including brick slips are produced in factory conditions and where alternative finishes altogether may be used.

22.4. The impact of automation on wellbeing

The full impact of automation and manufacturing-led construction on the wellbeing of workers is beyond the scope of this book, but it is a topic that requires a great deal of thought. The analysis presented suggests that there is likely to be a rapid technological change in the construction industry over the next decade. However, anyone who has been involved in significant change such as digitalisation will know that often the biggest challenge is managing the associated cultural change necessary to embed new technologies and ways of working. The Post Office Horizon IT scandal is a perfect example of poor implementation of technology and the real and significant consequences if it is not done well. So, to implement the automation and manufacturing-led technologies at the rate forecast, the cultural shift needs to be handled with care.

Research by the Institute for the Future of Work (Soffia *et al.*, 2024) identifies that the impact on the quality of life of workers of digital and communication technologies can be a positive one, likely because they are understood to be tools to assist in carrying out duties. However, the same research suggests that the introduction of newer, more advanced technologies can have a negative impact. This may be due to the potential complexity of new technologies or the concern that technologies may replace human jobs. The author has previously worked on programmes introducing new technologies such as the automation of compliance checking, where there has understandably been concern by those working in the field that their jobs would no longer exist, so there has been significant resistance to assisting the programme. This is a good example of the need to carefully explain the aims and implications of the introduction of technology because, as has already been stated, automation is generally aimed at improving productivity and doing more with the same resource, not reducing headcount.

With the changes forecast over the next decade, it is crucial that workers and other stakeholders are engaged in the process and their future opportunities clearly identified from the start. The analysis has shown that in the majority of cases, automation technologies are likely to work alongside humans and will change their focus to more creative, knowledge-based activities. Roles will typically require more of a technology focus than previously, however; for example, it is likely that site workers will more commonly use tablets or mobile devices to view drawings, requirements and tasks instead of using paper copies, and supervisors may rely more on helmet cameras or drones instead of walking sites every time they need to view progress.

It is likely that to take advantage of the automation and manufacturing-led technologies covered in this book, the workforce will need additional training. Training is one area that the construction industry has often struggled to manage; with such a high percentage of businesses being very small enterprises, it can be difficult to have the time and resources to invest in training staff, especially given the risk that they may then leave and any advantage could be lost. It is likely that to upskill the workforce to work alongside technologies that make the whole industry more productive, there will need to be elements of training provided either by Government, large partners within the supply chain (for example, integrators or manufacturers) or even by technology providers themselves. In the short to medium term, it is forecast that those with advanced skills who are capable of using new technologies will be able to command higher wages, so for individuals who have the opportunity to have training, it is likely to be beneficial. In the longer term there is likely to be a larger spread of wages as a result of new technology application, and there will be new occupations that we are currently unaware of.

For businesses such as architects or other consultants, there may be a need to change the way that fees are calculated. If tasks are automated and so take considerably less time, using current time and materials models the fees chargeable may become unsustainable. To resolve this, there may be a shift to a more value-based charging model. Value-based pricing has the advantage of allowing clients to pay for the consultant's expertise and the value they bring to an intervention rather than on time spent alone. This model encourages consultants to focus on delivering value for their clients, but defining the value and its cost can be a challenge.

22.5. Conclusion

The analysis has shown that there is significant potential for automation and manufacturing-led technologies to change the way the built environment is designed, delivered and operated over the next decade, but there is a limit to what can be achieved by only automating existing tasks instead of developing new tasks that take advantage of available technologies and resources. This is the reason why the book has described a process which starts by clearly defining what is needed, and then looking at how that can be systemised, and then automated and optimised. If the definition and systemisation stages are skipped, it is likely that the wrong things will be automated, jumping to a solution before understanding whether it is the right solution. In considering the development of new tasks to work alongside new technologies and resources, it is recommended that the Five Ps model described in Chapter 3 or a similar model is used in order to strike the right balance between technologies and other resources to deliver the required result.

The previous decade has seen quite a shift in many parts of the industry to more digitally enabled ways of working through BIM and the use of accurate laser scans, but the decade ahead is likely to see more widespread and rapid change still, not just in the technologies that will become available and be used, but also in the tasks that are carried out to take full advantage of these new resources to deliver the best results.

The level of automation of existing tasks varies quite considerably, with tasks involving recording and processing data much more likely to be automated than physical activities or those requiring the application of expertise. It is therefore likely that manufacturing-led solutions will be developed that minimise the need for physical work outside of the factory that cannot be automated but, as with automation technologies, this needs to be carefully considered from the outset as a complete set of activities, not simply automating some and assuming that remaining tasks will be carried out traditionally if there are better alternatives available; a whole asset approach needs to be used.

The future of work in the built environment industries is bright, with more work required than the industry can currently deliver, so automation is more likely to support the workforce to deliver more with the same resource than it is to replace it. However, there is a real need for a great deal of retraining and a willingness to adapt to new ways of working. The workforce needs to be engaged in the process of transition for it to be successful, and to provide confidence that technology is there to support workers in delivering and operating a more efficient and effective built environment.

The level of automation and manufacturing-led construction varies depending on how the world develops in the next few years, but the next chapter presents three potential scenarios and their implications.

REFERENCES

Construction News (2024) Despite the setbacks, MMC and offsite construction are still growing. https://www.constructionnews.co.uk/sections/cn-intelligence/despite-the-setbacks-mmc-and-offsite-construction-are-still-growing-11-03-2024/ (accessed 10/04/2024).

Make UK (2023) *Who Will Be the Builders? Modular's Role in Solving the Housing Labour Crisis*. Make UK, London, UK.

ONS (Office for National Statistics) (2010) *Standard Occupational Classification 2010 Volume 1: Structure and Descriptions of Unit Groups*. Palgrave Macmillan, Basingstoke, UK.

ONS (2023) Construction Statistics Annual Tables 2022. https://www.ons.gov.uk/business industryandtrade/constructionindustry/datasets/constructionstatisticsannualtables (accessed 22/03/2024).

Soffia M, Leiva-Granados R, Zhou X and Skordis J (2024) *Does Technology Use Impact UK Workers' Quality of Life? A Report on Worker Wellbeing*. Institute for the Future of Work, London, UK.

WEF (World Economic Forum) (2023) *Future of Jobs Report 2023*. WEF, Geneva, Switzerland.

emerald PUBLISHING   ice

Steve Thompson
ISBN 978-1-83608-599-7
https://doi.org/10.1108/978-1-83608-598-020241026

Chapter 23
An automated construction industry: 2035 scenarios

23.1. Introduction

This chapter presents three future scenarios based on different levels of adoption of both automation and manufacturing-led construction technologies between now and 2035. The starting point in terms of manufacturing-led construction technologies is the assumption that in 2024 manufacturing-led construction is used to deliver 11% of the construction industry's output by value. This is based on analysis by Glenigan (Construction News, 2024) and by NBS (NBS, 2023) and consists of

- 3D modules – 23%
- 2D panels – 34%
- components – 43%.

The change in level of adoption over the period analysed varies by scenario and will be described in the following sections.

The level of adoption of the different automation technologies described in Chapter 15 and elsewhere is based on the analysis carried out for the book and described in Chapter 22 and the Appendix. This analysis suggests that in 2024, 10% of tasks within the wider built environment are currently automated. This is significantly below the 34% suggested globally in the World Economic Forum's report (WEF, 2023), but that figure resulted from a survey of large organisations across many industries. As has been discussed, construction is potentially harder to automate than many other industries due to the physical nature of work and variety of projects and locations; however, the difference is not as wide as many may expect. This is supported by analysis in the WEF report (WEF, 2023) which identifies that the average level of automation in infrastructure in 2023 was 1% higher than the average across all industries, and in 2027 will be 3% lower than average, suggesting a slowing of the level of automation over time relative to other industries. As with the level of adoption of manufacturing-led construction, the level of automation identified in this chapter is based on different scenarios and is described in the following sections.

The level and speed of change in level of adoption of automation is influenced by a number of factors. McKinsey (McKinsey Global Institute, 2017) suggest five key factors influencing the level of automation

- technical feasibility – can something technically be automated?
- cost of implementation – what is the cost of the technology?
- labour market dynamics – the availability, cost and continuity of required labour compared to the availability and cost of automation technologies

▨ economic benefits – what benefits can automation deliver?

▨ regulatory and social acceptance – will workers and wider society accept and trust automation technologies over human workers?

These five factors have been considered in forecasting the potential speed of adoption of automation in the three scenarios, not just technical feasibility alone.

23.2. Scenario 1

Scenario 1 represents the slowest level of adoption of both automation and manufacturing-led construction technologies.

For manufacturing-led technology adoption, the forecast increase is based on the same rate as has been experienced between 2017 and 2023 according to Glenigan (Construction News, 2024), which represents a 1% increase per year. For simplicity, it has been assumed that the relative adoption of 2D, 3D and components remains the same. Table 23.1 presents the resulting percentage levels of adoption, which have then been applied to the tasks analysed along with the automation technologies to calculate the overall levels of automation.

To analyse the level of automation technology, scenario 2 (the middle scenario) is based on the level of adoption presented in Figure 15.1 in Chapter 15. The level of adoption in scenario 1 is 10% below this level for each technology.

The results of scenario 1 are presented in Table 23.2, which identifies the level of automation of tasks between 2024 and 2035 based on the application of automation and manufacturing-led technologies to existing tasks.

The results show that there is forecast to be a rapid change in the level of automation up to 2031, and then the level is expected to level off, with a maximum level of automation of 30%. This limit can be justified by the complexity and variety of projects and, if the human effort that has been automated is refocused on similar activities on other projects, this can result in a potential increase in output of £28.7 bn by 2035.

Table 23.1 Scenario 1 level of adoption of manufacturing-led construction technologies

	2024	2025	2026	2027	2028	2029	2030	2031	2032	2033	2034	2035
3D	2.3%	2.5%	2.8%	3.0%	3.2%	3.5%	3.7%	3.9%	4.1%	4.4%	4.6%	4.8%
2D	3.4%	3.7%	4.1%	4.4%	4.8%	5.1%	5.4%	5.8%	6.1%	6.5%	6.8%	7.1%
Components	4.3%	4.7%	5.2%	5.6%	6.0%	6.5%	6.9%	7.3%	7.7%	8.2%	8.6%	9.0%

Table 23.2 Scenario 1 level of automation of existing tasks through adoption of technologies

	2024	2025	2026	2027	2028	2029	2030	2031	2032	2033	2034	2035
By task	10%	14%	18%	21%	22%	24%	26%	26%	26%	26%	27%	27%
By time	11%	16%	20%	24%	25%	27%	29%	30%	30%	30%	30%	30%

23.3. Scenario 2

Scenario 2, or the middle scenario, is based on an annual increase in adoption of manufacturing-led technologies of 8.5%, in line with analysis by Market Reports World (Building Specifier, 2024), and the percentages are presented in Table 23.3.

As described in Section 23.2, the level of adoption of automation technologies is based on those presented in Figure 15.1 in Chapter 15.

The results suggest a 32% level of automation of tasks by 2035, and are presented in Table 23.4. This can deliver an increase in construction industry output of £29.5 bn by 2035.

23.4. Scenario 3

Scenario 3 represents the most significant increase in level of adoption of both manufacturing-led and automation technologies. The adoption of manufacturing-led technologies is based on a 50% adoption across the construction industry by 2035, based on the split between different technologies described in Section 23.1. The levels of adoption are presented in Table 23.5. This scenario for manufacturing-led construction is seen as optimistic, but achievable if the industry moves quickly.

The level of automation in scenario 3 differs considerably from scenarios 1 and 2, and is based on the level of automation identified by the WEF (WEF, 2023) for relevant industry sectors (for example, infrastructure for construction-related activities, but also real estate and supply chain for relevant occupations). One of the most significant differences between the WEF figures and the analysis carried out in this report is the initial assessment of the percentage of tasks that are automated. In this scenario, it is assumed that by 2027, the relevant industries reach the level of automation suggested by the WEF. From that point forward, the percentage change in level of

Table 23.3 Scenario 2 level of adoption of manufacturing-led construction technologies

	2024	2025	2026	2027	2028	2029	2030	2031	2032	2033	2034	2035
3D	2.3%	2.5%	2.7%	3.0%	3.2%	3.5%	3.8%	4.1%	4.5%	4.9%	5.3%	5.7%
2D	3.4%	3.7%	4.1%	4.4%	4.8%	5.2%	5.6%	6.1%	6.6%	7.2%	7.8%	8.5%
Components	4.3%	4.7%	5.1%	5.6%	6.0%	6.6%	7.1%	7.7%	8.4%	9.1%	9.9%	10.7%

Table 23.4 Scenario 2 level of automation of existing tasks through adoption of technologies

	2024	2025	2026	2027	2028	2029	2030	2031	2032	2033	2034	2035
By task	11%	16%	19%	23%	24%	25%	27%	28%	28%	28%	29%	28%
By time	12%	18%	22%	26%	27%	28%	31%	32%	32%	32%	32%	32%

Table 23.5 Scenario 3 level of adoption of manufacturing-led construction technologies

	2024	2025	2026	2027	2028	2029	2030	2031	2032	2033	2034	2035
3D	2.3%	2.5%	3.4%	4.3%	5.2%	6.1%	7.0%	7.9%	8.8%	9.7%	10.6%	11.5%
2D	3.4%	3.7%	5.1%	6.4%	7.7%	9.0%	10.4%	11.7%	13.0%	14.3%	15.7%	17.0%
Components	4.3%	4.7%	6.4%	8.1%	9.8%	11.4%	13.1%	14.8%	16.5%	18.1%	19.8%	21.5%

automation is assumed to be similar to that in the other scenarios, leading to a levelling off of the total level of automation between 2030 and 2035. This approach leads to the average level of automation across occupations and tasks, as illustrated in Table 23.6.

This and the adoption of manufacturing-led technologies forecasts a 51% automation of tasks by 2035, with intermediate levels presented in Table 23.7. This is considerably above the other scenarios, with the potential for a £52.8 bn increase in output from 2024. The potential level of automation in scenario 3 is likely to be a real challenge to meet across the industry, and will involve the development of new tasks to work alongside automation and manufacturing-led construction, so not solely reliant on the automation of existing tasks. It is considered unlikely that this level of automation will be achieved in sectors where bespoke requirements are required, such as in the remodelling of existing assets, but it may be achieved in some new-build sectors.

23.5. Summary

The potential for automation of tasks within the built environment industries is significant, and the technologies (both manufacturing-led and automation) are expected to be available over the next decade. The three scenarios presented here are compared in Figure 23.1.

Each of the scenarios suggests that there is a limit to the level that existing processes can be automated. That is to be expected; those processes were developed based on existing technologies, labour and requirements. Moving forward, it is likely that with increased access to automation technologies that new, complementary processes are developed which will significantly increase the level of automation towards scenario 3. However, if the automation of tasks reaches the levels predicted in any of these scenarios, there will be a significant improvement in productivity within the built environment, along with the other benefits that automation and manufacturing-led construction can bring, as described elsewhere in this book.

To maximise productivity in construction and related industries, and to optimise the use of automation and manufacturing-led construction, it is advisable to look beyond automating existing tasks and to follow the steps identified in Figure 1.2 and the Five Ps model to optimise the use of resources. With careful consideration, technology can enable a more productive, sustainable built environment.

Table 23.6 Average level of automation of tasks in construction-related industries based on data from WEF (2023)

	2023	2024	2025	2026	2027	2028	2029	2030	2031	2032	2033	2034	2035
Average percentage of adoption across tasks	10.0%	17.8%	25.6%	33.3%	41.1%	42.7%	45.4%	49.4%	50.7%	51.0%	51.3%	52.0%	51.5%

Table 23.7 Scenario 3 level of automation of existing tasks through adoption of technologies

	2024	2025	2026	2027	2028	2029	2030	2031	2032	2033	2034	2035
By task	19%	26%	34%	43%	44%	47%	52%	53%	54%	54%	55%	55%
By time	19%	26%	35%	43%	45%	48%	52%	54%	54%	55%	56%	56%

Figure 23.1 Level of automation of existing tasks across each scenario (Author's own)

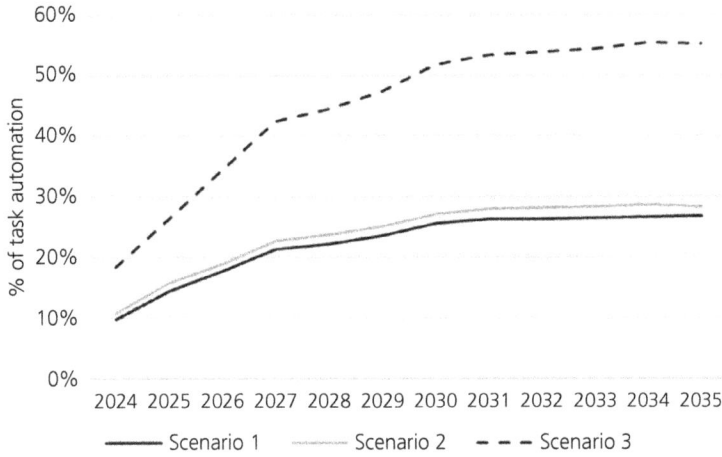

REFERENCES

Building Specifier (2024) Global growth in offsite construction. https://buildingspecifier.com/global-growth-in-offsite-construction/ (accessed 11/04/2024).

Construction News (2024) Despite the setbacks, MMC and offsite construction are still growing. https://www.constructionnews.co.uk/sections/cn-intelligence/despite-the-setbacks-mmc-and-offsite-construction-are-still-growing-11-03-2024/ (accessed 11/04/2024).

McKinsey Global Institute (2017) *Jobs Lost, Jobs Gained: Workforce Transitions in a Time of Automation*. McKinsey Global Institute, Dallas, TX, USA.

NBS (National Building Specification) (2023) *Digital Construction Report 2023*. NBS, Newcastle, UK.

WEF (World Economic Forum) (2023) *Future of Jobs Report 2023*. WEF, Geneva, Switzerland.

Steve Thompson
ISBN 978-1-83608-599-7
https://doi.org/10.1108/978-1-83608-598-020241027

Chapter 24
Conclusion

24.1. Introduction

There can be no doubt that the construction industry faces many challenges ahead and needs to adapt to meet them. The first few chapters of this book provided background on construction and related industries, and on the built environment that they deliver and maintain. The purpose was to understand the starting point for change, but also to see where lessons can be learned from the industry itself, and others that have transformed in recent decades. The construction industry has tried to change many times in the past – for example, with the Latham (1994) and Egan (1998) reports of the 1990s – but has struggled to shift from its existing model. So, why will the industry change this time? The short answer is because it needs to. Using Mark Farmer's (2016) rather blunt phrase, it may well be a case of modernise or die. Over the coming decades the industry needs to deliver significantly more assets which are better performing, with less resource (both human and material). It also needs to refurbish over 1 m homes per year in the UK, as well as other assets. So, existing delivery models just cannot cope with the scale of change in demand and expectations, but nor the pace of that change. To meet global and national carbon targets and demand for accommodation, the industry needs to adapt to deliver to these new levels quickly, not in the several decades that have historically been the case for transformation in other industries.

With the change that is needed comes great opportunity. Not only to deliver a more flexible, responsive and sustainable built environment, but also to be more productive. At the same time, from an individual perspective, it feels like an incredibly exciting time to be working in the industry, and one where traditional occupational boundaries are breaking down and offering the potential for new roles to work alongside technology. That leads on to another shift, which is an increasing focus on collaboration. The construction industry has at times (but not always) provided a rather confrontational and even hostile environment to work in, but that is beginning to change, and collaboration is becoming more widespread. Especially when facing rapid change, collaboration is hugely beneficial. If one part of the supply chain looks to change alone it can be incredibly difficult to land, whereas if organisations work together while considering the impact on those around them, change is more likely to stick for the benefit of all of those involved. The cultural change required should not be underestimated, but this is where learning from other industries such as software development can really help; businesses need to be more agile than has traditionally been the case, potentially with more than one business model to enable them to adapt to ever-changing requirements.

The construction industry today has a relatively small number of large businesses, with a significantly long tail of small- to medium-sized businesses, including the self-employed. Many industry initiatives in the past have focused too much on the smaller number of large players and not supported the majority of businesses within the industry. There is a real opportunity when moving to a more collaborative model to harness the long tail to deliver significantly more value, and technology can be one of the enablers for that to happen.

Another shift is to thinking about the whole life cycle of an asset, not just the delivery phase. This can not only deliver more value to the client but also provide new opportunities for future relationships and new business models.

One thing that the majority of technologies discussed in the book have in common is their reliance on data. Data comes from many sources, whether dynamic or static, but activities that can benefit the majority of technologies include structuring and managing data, whether that be product or service data, requirements or responses. Structuring data for a manufacturer can be a significant undertaking, but the good news is if it is done well, it only needs doing once and will add significant value to the manufacturer and customers, both now and in the future.

Moving on to project delivery or business model development for new technologies, there is a welcome move in many parts of the industry to focusing on outcomes, not outputs. In other words, defining what needs to be achieved, not what needs to be delivered. For this to work and deliver real value to customers and businesses alike, the first step is to clearly define what the desired outcomes or needs are. That should include future needs as well as those for today; an asset should not be considered as being finished once it has initially been built – that is only the beginning of its useful life – and so future flexibility and adaptability need to be considered to avoid unnecessary adaptations or demolitions at a later date. In defining requirements it is also important to not over-constrain delivery partners, which can unintentionally minimise value delivered by not taking into account the knowledge and experience of the supply chain. In defining need, it can be worth considering a platform approach if there is sufficient scale and volume, but at the very least it can be beneficial to understand the potential implications across a portfolio or a number of projects; this can be benefits to delivery partners using similar processes or products, or clients with repeatable requirements.

Once requirements have been clearly defined (whether for a business model or specific projects), systemisation of delivery can be considered. Systemisation before clearly defining a need can lead to inappropriate solutions being developed, or trying to squeeze systems into applications that don't really fit. When looking to systemise, it is important to consider the level of granularity – for example, is the system looking to vertically or horizontally integrate previous services, or provide a completely new delivery model? Either way, the Five Ps model described in Chapter 3 can be useful either as a simple checklist or as a more detailed model to assess the new approach and compare it with others. It can help identify gaps that need to be filled, whether through partnerships or additional services. Bearing in mind the need to be agile, a further consideration is the resources and capital needed to deliver a new systemised model. With the industry moving into a phase of constant change, it is likely that by the time new, significant production facilities are developed, not only demand but also competition will have already changed, and so a more asset-light approach as described in Chapters 4 and 14 may be sensible.

The book covers a wide range of automation technologies, the majority of which are equally applicable to traditional as well as manufacturing-led delivery models. The technologies are also applicable to all stages of an asset's life cycle, and many benefit from working alongside other technologies. When adopting new technologies to improve delivery, it is crucial that first the need is clearly defined to minimise the temptation to just apply the latest technology without good reason. Applying the wrong technology, or even the right technology at the wrong time, can be more disruptive than beneficial, so it is important to understand where a technology can help and how it impacts activities around it.

A common misconception when looking at automation is that it will replace human jobs. Hopefully this book has shown that it is more likely to increase jobs in the long run, but the majority of existing jobs will certainly change, and new ones will be created that haven't yet been thought of. As the research carried out for the book shows, there is a limit to the level of automation of existing tasks that can be achieved. However, as mentioned earlier, when applying new technologies or processes it is important to understand what changes around it. When technology provides new capabilities, the tasks that need to be carried out are likely to change, whether that is new tasks by humans or by other technologies. An example is generative design. Can computers be creative? Well, they can certainly deliver results that many would identify as being creative if developed by humans, but computers do not use the same creative processes to get there.

Hopefully this book has shown that the role automation and manufacturing-led construction should play in the future is to help deliver better outcomes, whether that means being more productive and delivering more with less, or whether it means being more agile and sustainable. In a way, the aim of both is to optimise assets and how they are delivered and maintained. A part of that involves considering the life cycle of both assets and their constituent parts. That can be designing assets for in-use adaptability to cope with changing use cases or designing systems to be demountable and reusable (not just recycled). To realise a sustainable built environment requires the full life cycle to be considered from the outset. For example, many of the buildings being built today will almost certainly require improvements to their structure or services in the next two decades if the UK is to meet its zero carbon targets by 2050, yet they are still being built because of the focus on short-term not long-term need. An alternative approach is to design out inflexibility and enable assets to be adaptable or easily upgradeable in the future without significant work. Part of that involves designing systems to be capable of being replaced, refurbished and to re-enter the supply chain for future use. This can potentially be achieved at no extra cost, just by changing the mindset from delivery focused on a point-in-time need to a life cycle benefit, and automation and manufacturing-led approaches can support delivery.

The future for construction and the built environment is bright. It is an exciting and rapidly changing industry, and has a strong base from which to adapt to meet the challenges ahead, but it needs to be more agile and open to new technologies and approaches. Hopefully this book has shown that change is already happening, and that the next few decades promise to be an exciting and fulfilling ride.

REFERENCES

Egan J (1998) *Rethinking Construction*. Department of Trade and Industry, London, UK.

Farmer M (2016) *The Farmer Review of the UK Construction Labour Model: Modernise or Die*. Cast Consultancy, London, UK.

Latham M (1994) *Constructing the Team*. The Stationery Office, London, UK.

emerald PUBLISHING   ice

Steve Thompson
ISBN 978-1-83608-599-7
https://doi.org/10.1108/978-1-83608-598-020241028

Appendix – Assessing the impact of automation

The purpose of this appendix is to describe the methodology used to provide quantitative estimates on the impacts of automation resulting from the application of technologies described throughout the book. It must be noted that the results are only generated to provide an indicative estimate across industry, not an accurate assessment of the potential impacts. The impacts are considered for each identified technology, both individually and in combination, at the level of individual tasks. As a result of this level of granularity, the assessments may be considered more accurate and specific than many pan-industry methodologies.

The starting point is to define what the boundaries of the industries to be considered are. As described in Chapter 3, for statistical purposes the construction industry is defined as much smaller than it actually is – for example, it excludes design teams and manufacturing businesses which most in the industry will recognise as being part of the industry. Table A1 identifies the industry classes that have been considered in this book as being part of the wider construction industry. The list is created from consideration of the full supply chain and professional services throughout the life cycle of a built asset, not purely the assembly stage of a construction project, with the purpose of providing a fuller picture on the potential automation of the industries who have a role to play in the built environment. The classes are taken from the Standard Industrial Classification (SIC) 2007.

Once the industry classes have been identified, all occupations which are employed within those industry classes have been identified. To do this, the Office for National Statistics (ONS) was asked to provide details of the number of people employed in the relevant industries by occupation in 2021. In addition to providing numbers of those employed, it provides information on the size of businesses that individuals were employed in, and a full list of the relevant occupations, described using the Standard Occupation Classification (SOC). For the purposes of the analysis, SOC 2010 was used instead of SOC 2020, as the data from ONS was returned in that classification. This revised list is combined with ONS data on the hours worked by occupation to later calculate the impact of automation in terms of monetary and duration values.

In the UK, the full SOC 2010 includes a list of activities that each occupation is likely to carry out as part of their usual duties. However, there is no information on the likely duration or frequency of those activities in a typical week. For this information, which is crucial to understand the real impact of a technology, o*net is used. O*net is a database of occupations and their relevant industries, tasks and activities in the USA. Because of the level of information included in o*net, the occupations identified in the UK as being relevant to the built environment need to be translated into occupations in o*net. Thankfully, there is an international occupations classification known as the International Standard Classification of Occupations (ISCO) 2008. The UK SOC 2010 occupations are therefore mapped to ISCO 2008, and then to o*net to enable the use of relevant o*net data to ascertain the potential impacts.

Table A1 Industry classes identified as being part of the wider construction industry for the purposes of analysis in this book (continued on next page)

SIC Code	SIC description
B	**MINING AND QUARRYING**
0811	Quarrying of ornamental and building stone, limestone, gypsum, chalk and slate
0812	Operation of gravel and sand pits; mining of clays and kaolin
C	**MANUFACTURING**
1610	Sawmilling and planing of wood
1622	Manufacture of assembled parquet floors
1623	Manufacture of other builders' carpentry and joinery
2311	Manufacture of flat glass
2331	Manufacture of ceramic tiles and flags
2332	Manufacture of bricks, tiles and construction products, in baked clay
2342	Manufacture of ceramic sanitary fixtures
2351	Manufacture of cement
2352	Manufacture of lime and plaster
2361	Manufacture of concrete products for construction purposes
2362	Manufacture of plaster products for construction purposes
2363	Manufacture of ready-mixed concrete
2364	Manufacture of mortars
2365	Manufacture of fibre cement
2369	Manufacture of other articles of concrete, plaster and cement
2370	Cutting, shaping and finishing of stone
2511	Manufacture of metal structures and parts of structures
2512	Manufacture of doors and windows of metal
2740	Manufacture of electric lighting equipment
2824	Manufacture of power-driven hand tools
3102	Manufacture of kitchen furniture
E	**WATER SUPPLY; SEWERAGE, WASTE MANAGEMENT AND REMEDIATION ACTIVITIES**
3900	Remediation activities and other waste management services
F	**CONSTRUCTION**
4110	Development of building projects
4120	Construction of residential and non-residential buildings
4211	Construction of roads and motorways
4212	Construction of railways and underground railways
4213	Construction of bridges and tunnels

Table A1 Continued

4221	Construction of utility projects for fluids
4222	Construction of utility projects for electricity and telecommunications
4291	Construction of water projects
4299	Construction of other civil engineering projects
4311	Demolition
4312	Site preparation
4313	Test drilling and boring
4321	Electrical installation
4322	Plumbing, heating and air-conditioning installation
4329	Other construction installation
4331	Plastering
4332	Joinery installation
4333	Floor and wall covering
4334	Painting and glazing
4339	Other building completion and finishing
4391	Roofing activities
4399	Other specialised construction activities
G	**WHOLESALE AND RETAIL TRADE; REPAIR OF MOTOR VEHICLES AND MOTORCYCLES**
4663	Wholesale of mining, construction and civil engineering machinery
4673	Wholesale of wood, construction materials and sanitary equipment
4674	Wholesale of hardware, plumbing and heating equipment and supplies
L	**REAL ESTATE ACTIVITIES**
6810	Buying and selling of own real estate
6820	Renting and operating of own or leased real estate
6831	Real estate agencies
6832	Management of real estate on a fee or contract basis
M	**PROFESSIONAL, SCIENTIFIC AND TECHNICAL ACTIVITIES**
7111	Architectural activities
7112	Engineering activities and related technical consultancy
N	**ADMINISTRATIVE AND SUPPORT SERVICE ACTIVITIES**
8110	Combined facilities support activities
8121	General cleaning of buildings
8122	Other building and industrial cleaning activities
8130	Landscape service activities

Once the list of UK occupations has been translated into o*net occupations, a full list of tasks carried out across all occupations is created. As stated earlier in this chapter, the purpose here is to provide a granular assessment of the impacts of automation, but there will certainly be tasks carried out in the delivery and operation of assets that are not included in the list created. However, across the 66 identified occupations, a total of 1398 tasks were assessed.

Each task in o*net includes information on the frequency of tasks carried out by each occupation and the relevant importance of each task. The frequency is identified against seven categories, outlined in the first two columns of Table A2.

For the purposes of this exercise, an estimation of the duration of a task based on its frequency was carried out and is identified in the final two columns of Table A2. This enables an estimation of the potential time saved through automation to be developed at a later stage.

Once data on the duration and frequency of all tasks has been assessed, a baseline is created for how many hours are spent on each task across all relevant employees. Then the potential for automation of each task as a result of a given technology is identified as a percentage chance. This enables the potential impact of each technology to be assessed across all occupations and employees. The potential impact can then be assessed of either individual technologies or combinations, asset life cycle stage or task type. Based on the average productivity by occupation, an indicative increase in output can be identified, in addition to time saved and tasks automated based on the level of adoption of technologies identified in Chapter 15. This analysis is presented in Chapter 22.

Finally, the technology adoption rates identified in each scenario in Chapter 23 can be used to provide an indication of the impact of level of adoption of each technology up to 2035, and from this the impact for each scenario can be identified. It must be noted, however, that such analysis does not include potential costs for the application of the technologies, nor does it include additional value that can be delivered beyond automating existing tasks.

Table A2 Assumed time taken for tasks based on frequency

Frequency category	Frequency description	Estimated duration of task	Estimated time spent on task per week (hours)
1	Yearly or less	8	0.2
2	More than yearly	8	0.4
3	More than monthly	2	1.2
4	More than weekly	1	3.0
5	Daily	1	5.0
6	Several times daily	0.5	7.5
7	Hourly or more	0.5	18.5

emerald PUBLISHING   ice

Steve Thompson
ISBN 978-1-83608-599-7
https://doi.org/10.1108/978-1-83608-598-020241029
Emerald Publishing Limited: All rights reserved

Index

Page numbers in *italic* indicate figures and tables.